Corrosion Protection of Metals and Alloys Using Graphene and Biopolymer Based Nanocomposites

T0199651

Editors

Hatem M.A. Amin

Cairo University
Giza, Egypt
and
Ruhr-University Bochum
Bochum, Germany

Ahmed Galal

Cairo University
Giza, Egypt

CRC Press
Taylor & Francis Group
Boca Raton London New York

CRC Press is an imprint of the
Taylor & Francis Group, an **Informa** business

A SCIENCE PUBLISHERS BOOK

Cover credit: Cover illustration reproduced by kind courtesy of the editors.

First edition published 2021
by CRC Press
6000 Broken Sound Parkway NW, Suite 300, Boca Raton, FL 33487-2742

and by CRC Press
2 Park Square, Milton Park, Abingdon, Oxon, OX14 4RN

© 2021 Taylor & Francis Group, LLC

CRC Press is an imprint of Taylor & Francis Group, LLC

Library of Congress Cataloging-in-Publication Data

```
Names: Amin, Hatem M.A., 1985- editor. | Galal, Ahmed, 1948- editor.
Title: Corrosion protection of metals and alloys using graphene and
    biopolymer based nanocomposites / editors, Hatem M.A. Amin, Cairo
    University, Giza, Egypt and Ruhr-University Bochum, Bochum, Germany,
    Ahmed Galal, Cairo University, Giza, Egypt.
Description: First. | Boca Raton : CRC Press is an imprint of Taylor &
    Francis Group, an Informa business, 2020. | "A Science Publishers Book."
    | Includes bibliographical references and index.
Identifiers: LCCN 2020021604 | ISBN 9781138046658 (hardcover)
Subjects: LCSH: Protective coatings. | Graphene--Industrial applications. |
    Biopolymers--Industrial applications. | Nanocomposites
    (Materials)--Industrial applications.
Classification: LCC TA418.76 .C675 2020 | DDC 620.1/1223--dc23
LC record available at https://lccn.loc.gov/2020021604
```

ISBN: 978-1-138-04665-8 (hbk)
ISBN: 978-0-367-64257-0 (pbk)

Typeset in Times New Roman
by Radiant Productions

Foreword

"This book provides a very accessible and useful guide to those who are keen to know the state-of-the-art of smart coatings. It covers comprehensively all aspects of the theoretical principles, characterization and application of carbon- and plant-based nanohybrids in corrosion protection. The book is written by recognized experts in their fields. It is recommended for researchers and professionals working in coating technology and corrosion science as well as an academic text for educational purposes."

Prof. Dr. Vinod K. Gupta

FWIF, FNASc, FRSC
Former Vice-Chancellor Dr. R. M. L. Avadh University, India
Distinguished Professor at King Abdulaziz University, Saudi Arabia
Thomson Highly Cited Researcher (h-index = 169 with more than 84270 citations)

"This book is a valuable book covering the recent advances in corrosion protection methods. The book mainly overviews advanced and environmentally friendly coatings, including graphene, polymers, and others and their monitoring and inspection methods. This book will be a valuable reference for those who are interested in corrosion protection, including professionals from the industrial sector as well as academic researchers in this field. The authors have made significant efforts to include recent topics in this book."

Prof. Dr. Mumtaz Quraishi

Chair Professor at Center of Research Excellence in Corrosion, KFUPM,
Saudi Arabia
50 years' experience in corrosion science

Preface

Corrosion causes huge economic losses for many industries. According to a NACE International study in 2016, the total global corrosion costs are estimated to be US $2.5 trillion per annum, which is equivalent to a significant part of the global Gross Domestic Product (GDP). Therefore, corrosion prevention is economically favourable compared to the expenses for the costly maintenance measures. In this context, we believe that reporting the most recent advances on corrosion control procedures from various research efforts is essential. Shedding the light on alternative, sustainable and non-classical scenarios for prevention/reduction of corrosion may help industry sectors to optimize the corrosion control practices. The authors of this book are academics or industrial experts in a particular field of corrosion.

This book reviews the latest results on various approaches towards sustainable protection of metals and alloys using green inhibitors instead of the harmful inorganic inhibitors. It also presents up-to-date reports on corrosion protection mechanisms using advanced and environmentally friendly coatings, including graphene, biopolymers and others. Moreover, it introduces the use of non-conventional and alternative inhibition approaches such as vapour phase- and plant-based inhibitors, self-healing smart coatings and plasma electrolytic oxidation. Furthermore, it demonstrates how to utilize various analytics to monitor and inspect the corrosion problems in many materials upon application of promising coatings. It also covers the theory and the current practices that help in understanding the corrosion processes and how to prevent or control corrosion, with numerous examples.

Among the versatile coatings available, graphene and biopolymer-based coatings represent current eco-friendly alternatives to conventional anti-corrosive coatings. Therefore, this book focuses on two main classes of materials which are introduced in two parts: Part one provides an overview of the application of biopolymers and plant-based materials as anticorrosive coatings. Part two looks at the use of graphene and its nanocomposites for corrosion control. The use of these materials showed efficient protection from corrosion while considering safety, health and environmental concerns. Therefore, this book highlights the current progress in the use of carbon-based and bio-polymeric materials for corrosion hindrance and other types of coatings are considered as well.

With contributions from experts from both academia and industry, this book is a valuable reference for those who are interested in the latest progress in corrosion inhibition approaches using green inhibitors and graphene nanocomposites, including professionals from the industrial sector as well as academic researchers in this field. Last but not least, if this book stimulates researchers to continue exploring optimal and sustainable solutions for the corrosion problem and helps industrial members to minimize the impact of corrosion on industry, it will have fulfilled its main objective.

Key Features*:*

- Overview of fundamentals, electrochemical aspects and types of corrosion processes of metals and alloys.
- Recent advances in coating materials with focus on graphene and bio-polymer-based coatings.
- Hands-on methods for corrosion inspection and characterization of anticorrosive films.
- Latest results on carbon and its allotropes as anticorrosive materials.
- Bio-polymer-based materials as an eco-friendly coating.
- Explores the potential of using nanomaterials for corrosion inhibition.
- Micro/nano-capsules for self-healing smart anticorrosion coatings.

April 2020
Bochum, Germany

Hatem M.A. Amin
Cairo University and Ruhr-University Bochum

Ahmed Galal
Cairo University

Acknowledgements

The editors would like to thank all the authors for their valuable contribution, successful cooperation and their interest. H. Amin wishes to thank his parents, brothers and friends for their continuous moral support. The authors acknowledge the continuous support from Cairo University.

Contents

Section 1
Corrosion Protection by Natural/Biopolymer Coatings

CHAPTER 1

Corrosion: Introduction

Anjali Peter[1] *and Sanjay K. Sharma*[2,*]

1. Introduction and History of Corrosion

Corrosion is as old as the earth itself and is known to people as rust. Corrosion is an undesirable phenomenon which destroys the lustre and beauty of the metal and lessens its life. Corrosion is the destruction and deterioration of metals as a result of reaction with the environment [1]. It is a major problem that must be confronted for safety, environmental and economic reasons in various chemical, mechanical, metallurgical, biochemical and medical engineering applications and more specifically in the design of a much more varied number of mechanical parts which equally vary in size, functionality and useful lifespan [2].

In 1834, *Bacquerel* projected that metal ion concentration differences are responsible for the corrosion process [3]. *De la Rive* indicated that the impurity was the main reason for deterioration [4]. Faraday's researchers [5] accumulated proof of the connection between chemical action and the electric currents. In 1847, Howe et al. revealed that differences in oxygen concentration in a rolling stream could give upsurge for a course of current between two parts of iron or zinc [6]. Developments in corrosion research changed promptly over the years. In 1950, application of polarization had been the topic of concern. From 1970 and later, corrosion research was focused on dissolution, localized corrosion and high temperature corrosion [7–10]. Nowadays, corrosion research is branching out into numerous study areas, likewise, the optical techniques have modernized the field. Surface analytical techniques, such as XPS [11, 12], XRD [13, 14], ESCA [15], SEM [16, 17], AES [18, 19], FTIR [20, 21] and AFM [22, 23], give us more valuable facts on the behavioural reactions, wideness, configuration, alignment of flicks and impact of surface oxide layers on the Detroit metals and alloys. Computers helped in the analysis of corrosion data such as impedance spectroscopy [24, 25]. The target objectives of all these studies are to minimize corrosion failures.

[1] Green Chemistry and Sustainability Research Group, Department of Chemistry, JECRC University, Jaipur-303905, India.

[2] Centre of Research Excellence in Corrosion, King Fahd University of Petroleum and Minerals, Dhahran 31261, Kingdom of Saudi Arabia.

* Corresponding author: sk.sharmaa@outlook.com

2. Cost and Effect of Corrosion

Increasing the rate of corrosion damage is a high hit to the country's economy. The monitoring expenditure of corrosion in industrialised body politic such as the U.S. and the European sum is about 3–5% of their gross national product [2]. The commercial budget of corrosion sphere is vast and walks into billions of dollars. It was estimated that one tonne of steel turns into rust every 90 seconds [26]. National Association of Corrosion Engineers–International India Section (NACE) reported that the annual global cost of corrosion in 2013 is estimated to be US$ 2.5 trillion which is equivalent to 3.4% of the global Gross Domestic Product (GDP) [27].

A number of unsafe effects of corrosion are as follows:

- Lessening of metal quality responsible for loss of mechanical forte and operational failure.
- Mishappening and harm to people arising from the operational partitioning (e.g., bridges, cars, and aircraft).
- Loss of valuable time in availability of industrial equipment.
- Condensed utility of things due to unattractiveness of exterior look.
- Contamination of liquids in containers and pipes (e.g., beer goes gloomy when minor amount of heavy alloy is getting mixed by corrosion).
- Loss in metallic properties like as lustre, malleability, texture, ductility and conductivity.

3. Principle of Corrosion

The Electrode Potential, Galvanic Series, Passivation and Polarization are the major terms which define the occurrence of corrosion, as depicted in Scheme 1.

Scheme 1. Principle of corrosion.

Principle of corrosion			
The electrode potential	**The galvanic series**	**Passivation**	**Polarization**
- The potential development by an electrode in equilibrium is a property of the metal forming the electrode. - The electrode potential is calculated under standard conditions with a pure metal as the electrode and an electrolyte containing unit concentration of ions of the same metal.	- Common Metals and alloys have tendency to corrode with surroundings. - This series arranges metals based on their reactivity order or qualitative scale.	- On increasing the potential of a metal electrode, the current density upsurges at first, after that it reaches a maximum value; it starts falling to a much lower value, then remains constant for further potential changes. This act is known as Passivation.	- The two electrode potentials tend to move towards each other due to the limited rates of the anodic and cathodic reactions. Differences between anode-cathode potential occurs with high current density, the electrode is said to be polarized, when current flows.

4. Theory of Corrosion

Ulick R. Evans (British scientist), who is considered by many as the "Father of Corrosion Science", has stated that "Corrosion is fundamentally an electrochemical process" [28]. Whereas corrosion can happen as one of the several mechanical bodies, the chemical mechanism of damage attack in aqueous environments will consist of a certain feature of electrochemistry. There will be a movement of electrons from specific areas of a surface in different areas through surroundings proficient in conducting ions. Whitney reported the worldwide satisfactory electrochemical theory [29]. Supplementary theories such as the acid theory, [30] chemical attack theory, colloidal theory [31] and the biological theory [11] were evidenced to form a portion of the electrochemical theory.

Electrochemical Theory

According to this theory, assortment or heterogeneousness on the metal surface is the key determinant to generate the galvanic cell which is essential for corrosion. This suggests that noble metals are corrosion free and there is no need to be divided as anodic and cathodic zones on the corroding metal. Corrosion is a redox chemical process, individually acquiring anodic and cathodic zones over the metal. These zones arise casually over the metal surface and they have a habit of swinging around on the whole surface triggering uniform corrosion. In brief, we can say corrosion occurs because of the thermodynamic instability. Hematite (Fe_2O_3) is reduced to produce iron in the method.

$$2\ Fe_2O_3\ (Hematite\ coke) + 3C \rightarrow 4\ Fe + 3\ CO_2 \uparrow$$

Steel, when faced with humid air, inclines to return to its native form of lower energy state.

$$4Fe + 3O_2 + n\ H_2O \rightarrow 2\ Fe_2O_3 \cdot nH_2O\ (Rust)$$

Anodic reaction:

$$Fe \rightarrow Fe^{+2} + 2e^-$$

The anodic reaction leads to the formation of Fe^{+2} ions.

Cathodic reaction: It can be either hydrogen evolution or oxygen reduction.

i) **Hydrogen Type:**

$2\ H_3O^+ + 2e^- \rightarrow H_2\uparrow + H_2O$ (acid solution in absence of O_2)

$2\ H_2O + 2e^- \rightarrow H_2 + 2\ OH^-$ (neutral and alkaline solution in absence of O_2)

ii) **Oxygen type:**

$2\ H_2O + O_2 + 4e^- \rightarrow 4\ OH^-$ (Neutral or alkaline, O_2 is the absorbed media in presence of O_2).

$4\ H^+ + O_2 + 4e^- \rightarrow 2\ H_2O$ (Acid medium in presence of O_2)

The oxidation handling of metal strikes at anodic region and lateral reduction took place at cathodic region. Due to the equal occurrence of both zone reactions, they

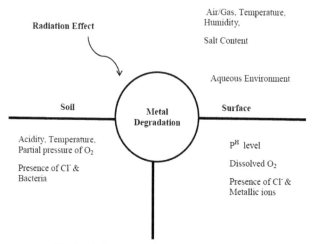

Fig. 1. Environmental degradation of a material.

fulfil the prior condition for corrosion. Corrosion processes come about in distinct areas on the metal covering due to the rough exterior of the metal surface or due to a fluctuation of concentration in the liquid phase.

5. Features of Corrosion

The following point represents the significant aspects that impact the corrosion process:

I. **Characteristics of the metal:** Sensitive metals such as Mg, K get corroded and noble metals (Ag, Au, Pt, and Pd) stay nonreactive with the surroundings. Some metals (Al, Cr, Ti, Ta, and Zn) show the passivity properties.

II. **Surrounding:** Surroundings comprise PH, humidity, temperature, impurities of the environment in which metal is exposed.

III. **The concentration of electrolyte:** Corrosion is an electrochemical phenomenon, meaning there is electron donation and receiving process and vice versa, so electrolytes act as a mediator to carry these electrons from one place to another, and hence the rate of corrosion is proportional to the concentration of electrolytes.

IV. **Temperature:** Some corrosion reactions take place at low temperature (reactive metal) while some others occur at high temperature (reactive and passive metal).

V. **Electrode potential:** Genesis of electrode potential happens because of polarisation, which lies in current flow. Polarisation occurs at the anode and is known as anodic polarisation or occurs at the cathode and is known as cathodic polarisation. Current flow provides the proof of anodic and cathodic polarisation, anodic potential decreases at the initial level and cathodic potential increases at the initial level. The potential becomes more reliable and efficient if it shifts to anodic polarisation (positive direction); because of the anodic polarisation the habit of creating oxide layer (oxidation process) decreases which condenses the rate of corrosion, but if it shifts to the negative side that's cathodic polarisation.

VI. **Hydrogen overvoltage:** This process connects with the oxidation-reduction processes. The vulnerability of corrosion is decided by hydrogen over voltage, and its lower value upsurges the corrosion process while its higher value decreases the corrosion rate [32].

VII. **Effect of impurities and conducting species:** Presence of NO_2, HCl and CO_2 acts as catalyst for providing favourable condition for corrosion. Conducting elements such as chemical salts upsurge the rate of corrosion; the higher the concentration of conducting species in environment, the higher the corrosion rate. In conducting medium, ions exchange process takes place very easily.

6. Types of Corrosion

Corrosion expresses its presence in different forms. It is important to know the type of corrosion taking place. For identical confirmation, it is essential to know the different types and characteristics of corrosion. Detailed forms of corrosion are mentioned below in Table 1 [33].

Table 1. Types of corrosion.

General corrosion	Localized corrosion	Metallurgic alloy influenced corrosion	Mechanically assisted degradation	Environmentally induced cracking
Equally damaged surface phenomena or uniform distribution of corrosion	Specific sites of metal surface degrade highly	Deterioration of metal caused by alloying or heat treatment	Depreciation in mechanical goods	Crack, split or narrow space occurrence between two surfaces visible with the attendance of force or stress
Atmospheric Corrosion Galvanic Corrosion Stray-current Corrosion General Biological Corrosion	Crevice corrosion Pitting corrosion Filly-form corrosion	Dealloying corrosion Inter-granular corrosion	Corrosion fatigue Civilization Fretting corrosion Interfacial corrosion	Hydrogen damage Stress corrosion Cracking corrosion

6.1 General Corrosion

This usually appears at the facade and superficies uniformly. In uniform corrosion, the corrosive surrounding occupies all parts of the metal surface and the metal composition must be uniform [34]. It is one of the most common corrosion forms that is responsible for the reduction in quality of material.

Example: *Steel in acid solutions.*

6.2 Intergranular Corrosion

This type mostly occurs in alloying objects; the main cause of this corrosion is the presence of galvanic elements. In this process, in alloying objects, there is an area of reactive metal (grain boundaries) which behaves as an anode, while other area acts as the cathode. The exposed surface area between anode and cathode is large and it enhances the possibilities of high corrosion rate. This process is spontaneous, diffuses deep into the metal, resulting in harm of productivity and sometimes severe failures.

6.3 Pitting Corrosion

The metal is consumed away and intruded in domiciles in the mode of holes, and the other part of the exterior changes. It is known as pitting corrosion. During pitting corrosion, the entire metal surface turns into rough and irregular in shape. In some other cases, the pits have occurred in certain metal surface areas, inflicting no harm to another part of the metal surface. Pitting corrosion also accumulates crevice corrosion, water-line attack and erosion corrosion attack [35].

6.4 Exfoliation

The peeling of layers from a solid metal alloy is called exfoliation. This appears in distorted objects. Corrosion supports "fibre orientation". This type of dissolution is generally perceived in wrought products which exhibit elongated structures. It is a sort of intergranular corrosion where surface grains from corrosion products are created at the grain boundaries under the surface.

6.5 Dealloying or Selective Corrosion

Corrosion outbreak takes place in structural elements. Dealloying is appearing in alloys where one component is comparatively hyperactive (less noble) than other elements. The hyperactive metal leaves the place in the form of pores, and sometimes these pores are covered with other corrosion products.

6.6 Galvanic Corrosion

This process happens in the electrically conductive environment. Deformation comes into existence due to the potential difference between the metals. If the active noble metal acts as the cathode with higher surface area and the less noble metal acts as the anode with less surface area, in this case, a large cathodic reaction takes place with the combination of higher anodic reaction. This process upsurges the corrosion rate with major harm to metal. The electromagnetic series (Galvanic series) is used to predict the tendency of metal to galvanic corrosion.
Example: *Aluminium and steel alloys.*

6.7 Water Line Attack

This phenomenon is often observed at water-air interfaces. Development of oxygen concentration lockups is mainly accountable for this type of attack. An example is a steel pipe driven into the soil which is particularly attacked just below the ground water level. A steel pipe dipped in sea water is most strongly attacked just below the water line. Protective coatings are used to reduce this type of attack.

6.8 *Filli-form Corrosion*

This form is easy to treat when it is in an early phase; this corrosion type appears on metal surface with organic coatings in the presence of 78–80% humidity. Filli-form corrosion occurrence is like a worm or scratches undercoating.
Example: *Oil and petroleum fields.*

6.9 *Stress Corrosion Cracking*

Combination of corrosion and mechanical stress leads to stress corrosion. It appears in the form of cracks at stress level below the tensile strength of metal [33].

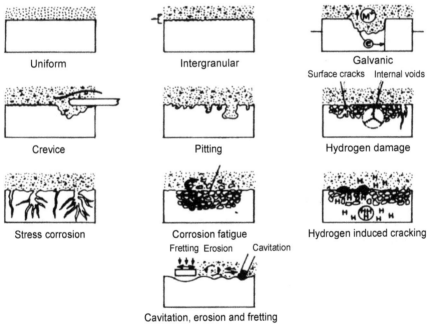

Fig. 2. Types of corrosion.

7. Thermodynamics of Corrosion

The study of the correlations between temperature and energy, work and properties of matter is recognised as thermodynamics, and it is fundamental to materials science. As shown in Fig. 3, energy contribution in positive direction indicates stability for metal; in materials, a natural origin of such energy is the thermal energy. Temperature increase plays a part in the activation for atoms vibrations with high amplitude, thus causing translational movement of atoms and the associated energy with the

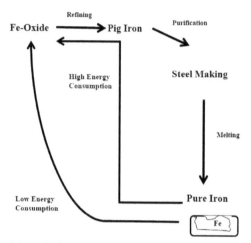

Fig. 3. Schematic diagram of the thermodynamic process of corrosion.

fluctuations and rotational translation assisting in taking a material from a meta-stable state to a stable state. This process makes use of excessive energy. Probably, if a metal (Iron) block is put in the ordinary real situation, after some interval it would turn back to its meta-stable condition either by degradation or "corrosion". The energy spent in this process is quite less.

8. Kinetics of Corrosion

A chemical reaction involves separation of bonds in reacting molecules and formation of new bonds in products. The characteristics of bonds are dissimilar in several substances and the rates of chemical reactions differ a lot from one another.

The rate of reaction depends upon constitutional and energetic factors which are merely defined by the thermodynamic quantities such as free energy change. Chemical kinetics is a complementary approach to thermodynamics for investigating a reaction.

The following major components are inherent to sustain the mechanism of the reaction:

I.   *Anode (Oxidation Site)*
II. *Cathode (Reduction Site)*
III. *Electrolytes (Ions)*
IV. *Conductor (to bear charges from one place to another).*

As per Fig. 4, when a metal is placed into aqueous surroundings, the electrolytes act with the metal exterior and exchange their charges with the help of conductors. The oxidation takes place at the anodic site, liberated electrons are accepted at the cathodic site and the reduction occurs as corrosion. Hydrogen gas is evolved during this process.

According to the law of mass action, the rate of a reaction depends on the molar concentrations (masses) of reactants and concentration keeps on decreasing with time. Thus, one can plot the concentration–time graph at the point corresponding to

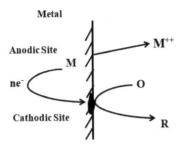

Fig. 4. Schematic of the kinetic process of corrosion.

the time (t) and get the slope which assures the rate of reaction. Kinetics helps us to find out the order of a reaction and it is an experimental quantity.

9. Corrosion Inhibition

An inhibitor is a species that creeps down or lessens a chemical reaction. A corrosion inhibitor is a substance which, when combined with an atmosphere, reduces the rapidity of attack by the corroding vicinities. The aim of corrosion inhibitors becomes one of the prime methods of encountering corrosion. To aid in finding efficient inhibitors, three points must be taken into consideration [19]:

1. Identification of decomposition complications.
2. The commercial position of the reservation process: It should be distinguished whether the corrosion damage goes beyond the price of the inhibitor and the maintenance of the subordinate manual procedure or not.
3. The adaptability of the inhibitor with the course of action being dealt with: This must be clear to keep away from unfortunate consequences.

Corrosion Inhibitors:

Again, corrosion inhibitor is an element which, when supplemented in minimal frequencies to a situation, decreases or prevents corrosion [36]. Over the ages, considerable attempts have been extended to identify useful corrosion inhibitors of an essential originating spot in numerous destructive media [37]. In corrosive media, nitrogen-based components and their secondaries, sulphur-holding complexes, aldehydes, thio-aldehydes, acetylenic composites and several alkaloids such as palavering, strychnine, quinine and nicotine inhibitors are effective toward corrosion inhibition. In neutral media, benzoate, nitrate, chromate, and phosphate behave as positive inhibitors. Inhibitors restrain or restrict the response of the metal with the media [38].

9.1 Synthetic Inhibitors

Artificial inhibitors are incorporated into the research facility to function as an adequate substitute for the original entities. Various synthetic corrosion inhibitors that have been operative across all corporations are reviewed in Table 2 [39].

Table 2. Examples of typical synthetic inhibitors.

Synthetic corrosion inhibitor	Common structure	Commercial name	Use
Benzyl Chloride Quaternary compounds		Alpha 1018, 1038, 1458, 1505, 3013, 3444	Used to formulate corrosion preventives
Diethylsulfate Quaternary compounds		Alpha 1080	Used to make corrosion preventives
Amine Ethoxylates Compounds		C1815	Used in oil and water-soluble solutions
Imidazoline compounds		Alpha 1153,1158	Used to formulate corrosion inhibitors
Organic boron compounds		Alpha 3220	Used to formulate corrosion inhibitor for oxygen, carbon dioxide, hydrogen sulphide, mineral acid and dissolved salts.

9.2 Green Inhibitors

Green corrosion inhibitors are eco-friendly and do not comprise excessive elements or other deleterious mixtures. An inhibitor is a raw material (or a variety of elements) summed up in a somewhat small intensity to deal with the exterior of a metal that brought to light a corrosive condition that eliminates or shrinks the decay of an element. These are known as spot closing materials, sinking variety or adsorption site blockers, owing to their adsorptive properties [40, 41]. The expression "*Green inhibitor*" or "*Eco-friendly inhibitor*" has a relation to the materials that have biocompatibility in the real atmosphere. Various green corrosion inhibitors used for corrosion inhibition of various metals are tabulated in Table 3.

In this book, the focus is on the inhibition using advanced environmentally friendly coatings including graphene and biopolymers. In particular, the book covers

Table 3. Examples of green corrosion inhibitors.

Metal	Inhibitor source	Active ingredient
Steel	Tea leaves	---
Steel	Pomegranate juice and Peels	---
Steel	Emblica officinalis	---
Steel	Terminalia bellerica	---
Steel	Eucalyptus oil	Monomtrene 1,8-cineole
Steel	Rosemary	
C-steel, Ni, Zn	Lawsonia extract (Henna)	Lawsone (2-hydroxy-1,4-napthoquinone resin and tannin, coumarin, Gallic, acid, and sterols)
Mild steel	Garcinia kola seed	Primary and secondary amines, unsaturated fatty acids and Biflavnone
Steel	Mango/orange peels	---
Steel	*Hibiscus sabdariffa (Calyx extract)* in 1M H_2SO_4 and 2M HCl solutions, Stock 10–50%	Molecular protonated organic species in the extract. Ascorbic acid, amino acids, flavonoids, pigments and carotene
Al	*Opuntia* (modified stems cladodes)	Polysaccharide (mucilages and pectin)
Al, steel	Aqueous extract of tobacco plant and its parts	Nicotine
Al-Mg alloy	Aqueous extract of Rosmarinus officinalis—Neutral phenol subfraction of the aqueous extract	Catechin
Al	*Prosopiscineraria* (khejari)	---
Al	Tannin beetroot	---
Al	Saponin	---

the recent advances in research and development in this growing field in order to minimize the impact of corrosion on economy.

10. Concluding Remarks

Corrosion is all about a destruction phenomenon for metals. A small initiation of degradation of material leads to big breakdown and it directly affects the economy of a nation. Indeed, it is possible to minimise the effect of corrosion by using corrosion inhibitors. Synthetic corrosion inhibitors are applicable on a large scale, but their high cost and maintenance pressurised human brain to search for alternative inhibitors, those that would be eco-friendly. Green inhibitors with known active compound plants could be the most promising coating. Nowadays, it is becoming a trend to use green inhibitors due to the ready availability of several plants. The science in the ancient world was very advanced as people knew the medicinal properties of many plants many years ago. Science always leads the path of unknown things in search to get a new thing and make it known to the world and finally use it for human benefits.

References

[1] Alavi, A. and R. A. Cottis. 1987. The determination of pH, potential and chloride concentration in corroding crevices on 304 stainless steel and 7475 aluminium alloy. Corrosion Science 27: 443–451.

[2] Bhaskaran, R., N. Palaniswamy, N. S. Rengaswamy and M. Jayachandran. 2005. Global cost of corrosion—a historical review. pp. 621–628. *In*: Cramer, S.D. and B.S. Covino (eds.). Corrosion: Materials, ASM International.

[3] Becquerel, A. C. 1834. Experimental treatise on electricity and magnetism, and their relations with natural phenomena. Vol. Didot.

[4] De La Rive, A. 1830. "N"ote relatire a l'action qu'exerce sur le zinc l'acide sulfuriọue etendu d'eau. Ann. Chira. Phys. (2. ser.), 43: 425–440.

[5] Tweney, R. D. 1992. Inventing the field: Michael Faraday and the creative of electromagnetic field theory. Inventive Minds: Creativity in Technology 31.

[6] Bibliography of the metals of the platinum group: Platinum, palladium, iridium, rhodium, osmium, ruthenium, 1748–1917, 1919.

[7] Gaiser, L. and K. E. Heusler. 1970. Die kinetik der zinkelektrode in zinkperchloratlösungen, Electrochimica Acta 15: 161–171.

[8] Heusler, K. E. and L. Gaiser. 1970. The mechanism of the cadmium electrode. Journal of the Electrochemical Society 117: 762.

[9] 'M. Bockris, J. O. 1989. Spectroscopic observations on the nature of passivity. Corrosion Science 29: 291–312.

[10] Wu, X. Q., Y. S. Yang, Q. Zhan and Z. Q. Hu. 1998. Structure degradation of 25Cr35Ni heat-resistant tube associated with surface coking and internal carburization. Journal of Materials Engineering and Performance 7: 667–672.

[11] El Azhar, M., M. Traisnel, B. Mernari, L. Gengembre, F. Bentiss and M. Lagrenée. 2002. Electrochemical and XPS studies of 2,5-bis(n-pyridyl)-1,3,4-thiadiazoles adsorption on mild steel in perchloric acid solution. Applied Surface Science 185: 197–205.

[12] Hermas, A. A., M. Nakayama and K. Ogura. 2005. Formation of stable passive film on stainless steel by electrochemical deposition of polypyrrole, Electrochimica Acta 50: 3640–3647.

[13] Rajendran, A. 2011. Isolation, characterization, pharmacological and corrosion inhibition studies of flavonoids obtained from Nerium oleander and Tecoma stans. International Journal of PharmTech Research 3: 1005–1013.

[14] Yeh, J. -M., C. -L. Chen, Y. -C. Chen, C. -Y. Ma, K. -R. Lee, Y. Wei and S. Li. 2002. Enhancement of corrosion protection effect of poly(o-ethoxyaniline) via the formation of poly(o-ethoxyaniline)–clay nanocomposite materials. Polymer 43: 2729–2736.

[15] Woodruff, D. P. 2016. Modern Techniques of Surface Science. Cambridge University Press, Cambridge.

[16] Migahed, M. A., M. Abd-El-Raouf, A. M. Al-Sabagh and H. M. Abd-El-Bary. 2005. Effectiveness of some non ionic surfactants as corrosion inhibitors for carbon steel pipelines in oil fields. Electrochimica Acta 50: 4683–4689.

[17] Ellmer, K. 2000. Magnetron sputtering of transparent conductive zinc oxide: relation between the sputtering parameters and the electronic properties. Journal of Physics D: Applied Physics 33: R17–R32.

[18] Quraishi, M. A., J. Rawat and M. Ajmal. 2000. Dithiobiurets: a novel class of acid corrosion inhibitors for mild steel. Journal of Applied Electrochemistry 30: 745–751.

[19] Zhang, X., W. G. Sloof, A. Hovestad, E. P. M. van Westing, H. Terryn and J. H. W. de Wit. 2005. Characterization of chromate conversion coatings on zinc using XPS and SKPFM. Surface and Coatings Technology 197: 168–176.

[20] Amalraj, A., M. Sundaravadivelu, A. P. Regis and S. Rajendran. 2001. Corrosion inhibitor for carbon steel in ground water. Bulletin of Electrochemistry 17: 179–182.
[21] Ravichandran, R., S. Nanjundan and N. Rajendran. 2004. Effect of benzotriazole derivatives on the corrosion of brass in NaCl solutions. Applied Surface Science 236: 241–250.
[22] Alves, V., A. M. C. Paquim, A. Cavaleiro and C. Brett. 2005. The nanostructure and microstructure of steels: Electrochemical Tafel behaviour and atomic force microscopy. Corrosion Science—CORROS SCI. 47: 2871–2882.
[23] Blunt, J. W., B. R. Copp, R. A. Keyzers, M. H. G. Munro and M. R. Prinsep. 2016. Marine natural products. Natural Product Reports 33: 382–431.
[24] Adey, R. A., S. M. Niku, C. A. Brebbia and J. Finnegan. 1986. Computer aided design of cathodic protection systems. Applied Ocean Research 8: 209–222.
[25] Kendig, M. W., E. M. Meyer, G. Lindberg and F. Mansfeld. 1983. A computer analysis of electrochemical impedance data. Corrosion Science 23: 1007–1015.
[26] Goldie, B. P. and J. J. McCarroll. 1984. Inhibiting corrosion in aqueous systems. US Patent.
[27] Jacobson, G. 2016. International Measures of Prevention, Application, and Economics of Corrosion Technologies Study. Houston, USA.
[28] Evans, U. R. 1981. An Introduction to Metallic Corrosion. 3rd Ed., Edward Arnold, London.
[29] Whitney, W. R. 1903. The corrosion of iron. J. Am. Chem. Soc. 25: 394–406.
[30] Walker, W. 1913. The corrosion of iron and steel. J. Ind. Eng. Chem. 5: 444–445.
[31] Tremont, R., H. de Jesús-Cardona, J. García-Orozco, R. J. Castro and C. R. Cabrera. 2000. 3-Mercaptopropyltrimethoxysilane as a Cu corrosion inhibitor in KCl solution. Journal of Applied Electrochemistry 30: 737–743.
[32] Kucera, V. (ed.). 1988. The Effect of Acidification on Corrosion of Structures and Cultural Property. John Wiley & Sons Ltd.
[33] Lynes, W. 1951. Some historical developments relating to corrosion. J. Electrochem. Soc. 98: 3C.
[34] Yamashita, M., H. Konishi, T. Kozakura, J. Mizuki and H. Uchida. 2005. *In situ* observation of initial rust formation process on carbon steel under Na_2SO_4 and NaCl solution films with wet/dry cycles using synchrotron radiation X-rays. Corrosion Science 47: 2492–2498.
[35] Leyerzapf, H. 1985. Zur Bedeutung und Geschichte des Wortes "Korrosion". Werkstoffe und Korrosion 88–96.
[36] Al-Otaibi, M. S., A. M. Al-Mayouf, M. Khan, A. A. Mousa, S. A. Al-Mazroa and H. Z. Alkhathlan. 2014. Corrosion inhibitory action of some plant extracts on the corrosion of mild steel in acidic media. Arabian Journal of Chemistry 7: 340–346.
[37] Ebenso, E. E., M. M. Kabanda, T. Arslan, M. Saracoglu, F. Kandemirli, L. C. Murulana, A. K. Singh, S. K. Shukla, B. Hammouti, K. F. Khaled and M. A. Quraishi. 2012. Quantum chemical investigations on quinoline derivatives as effective corrosion inhibitors for mild steel in acidic medium. International Journal of Electrochemical Science 5643–5676.
[38] Fiala, A., A. Chibani, A. Darchen, A. Boulkamh and K. Djebbar. 2007. Investigations of the inhibition of copper corrosion in nitric acid solutions by ketene dithioacetal derivatives. Applied Surface Science 253: 9347–9356.
[39] Peter, A., I. B. Obot and S. K. Sharma. 2015. Use of natural gums as green corrosion inhibitors: an overview. Int. J. Ind. Chem. 6: 153–164.
[40] Jerkiewicz, G. 1995. Examination of factors influencing promotion of H absorption into metals by site-blocking elements. J. Electrochem. Soc. 142: 3755.
[41] Henríquez-Román, J. H., L. Padilla-Campos, M. A. Páez, J. H. Zagal, M. A. Rubio, C. M. Rangel, J. Costamagna and G. Cárdenas-Jirón. 2005. The influence of aniline and its derivatives on the corrosion behaviour of copper in acid solution: a theoretical approach. Journal of Molecular Structure: Theochem. 757: 1–7.

CHAPTER 2

Biopolymer Composites and Nanocomposites for Corrosion Protection of Industrial Metal Substrates

*Saviour A. Umoren** and *Moses M. Solomon*

1. Introduction

Corrosion, commonly defined as the deterioration of a material (usually a metal) or its properties because of a reaction with its environment, is a global problem. NACE International, The Corrosion Society, estimates that global corrosion and its consequences cost developed nations about 3–5% of GDP or GNP [1]. Methods commonly employed to combat corrosion include cathodic protection, materials selection, coatings and linings and corrosion inhibitors. Corrosion inhibitors form a layer over the metallic substrate and protect the metal from corrosion, thereby enhancing the life of the metal. Coatings designed for corrosion protection must offer an effective physical barrier, impeding the access of aggressive species to the metallic interface. For many years, the most effective corrosion protection systems were based on the use of chromate-rich surface treatments and/or primers and pigments based on chromates [2]. However, the current legislation imposed by REACH (Registration, Evaluation, Authorization and Restriction of Chemicals) prohibits the use of hexavalent chromium in almost all sectors except the aerospace industry [3]. Many alternatives have been explored so far, including a wide range of "green" surface treatments and pretreatments, environmentally safe pigments and natural corrosion inhibitors [4]. The latest developments propose coatings with low volatile organic compounds (VOCs), based on waterborne formulations and isocyanate-free compositions, as well as smart and self-healing polymers [4].

Centre of Research Excellence in Corrosion, Research Institute, King Fahd University of Petroleum and Minerals, Dhahran 31261, Kingdom of Saudi Arabia.
* Corresponding author: umoren@kfupm.edu.sa

The major drawbacks of conventional coatings include high risk of cracking and failure due to poor bonding between the coatings' materials and metal substrates [5]. One of the most promising alternatives to overcome the drawbacks relies on the use of biocompatible polymeric or composite coatings designed to decrease the corrosion resistance to acceptable levels [6]. The improvement of the adhesion, mechanical properties and barrier effect of polymer coatings are necessary to enhance their efficiency in corrosion protection. It has been reported that the adhesion, mechanical, and barrier properties of polymer can be increased using organic and inorganic fillers [7, 8]. Furthermore, nano scaled fillers imply better barrier properties in polymer coatings, even at low concentrations, compared to the micron sized additives in the matrix of polymer coatings [7, 8].

Biopolymers are naturally derived polymeric biomolecules which could, on the basis of their monomeric unit composition, be grouped into three classes: polysaccharides; protein/polypeptides; and polynucleotides [9]. Cellulose, chitosan, and alginate are the common polysaccharides. Silk, collagen, and keratin are the typical examples of protein. An example of polynucleotide is deoxyribonucleic acid (DNA). These biopolymers are derived from natural sources like plants, exoskeletons of arthropods, skin, silkworm cocoon, spider webbing, hair, etc. [10–12] as illustrated in Fig. 1. They are renewable, biocompatible, biodegradable, flexible, and possess multiple reactive sites hence attractive in diverse areas of human endeavors [9]. Generally, both naturally-occurring and synthetic polymers have been tested as metal corrosion inhibitors with the intention to replace the toxic inorganic and organic corrosion inhibitors with them [4]. However, most of them only behaved as

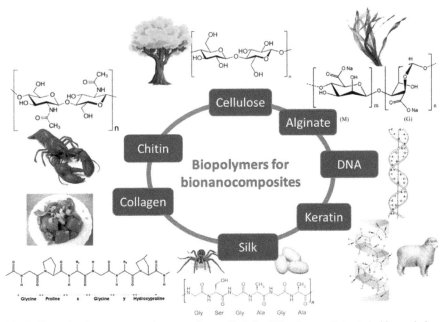

Fig. 1. The molecular structures and sources of naturally derived biopolymers. Extracted with permission from Xiong et al. [9]. Copyright 2018 Elsevier B.V.

moderate corrosion inhibitors [4, 13]. This observation is the motivation for several attempts such as copolymerizing [14], addition of substances that exert synergistic effect [14–17], cross linking [17, 19], blending [20], and most recently incorporation of inorganic substances in nano size into the polymer matrix [21–23] which have been made to improve the inhibition ability of polymers. The application of polymer composites and nanocomposites as anticorrosion materials have shown promising results and are believed to form metal chelate which could barricade metal surfaces from corrosive agents [21–23]. In this chapter, the application of biopolymer composites and nanocomposites for corrosion protection of different metal substrates in different corrosive media is highlighted.

2. Biopolymer Composites and Nanocomposites for Corrosion Inhibition

As the quest of developing metals' corrosion inhibitors that are not just effective but also friendly with the natural environment intensifies, biopolymer composites and nanocomposites are increasingly tested as corrosion inhibitors for various substrates in diverse kinds of aggressive media. Compositing is a modification technique that involves the reaction(s) of different components to produce a material that is chemically distinct with characteristics that are different from the properties of the individual components. For a nanocomposite, the dimension of one of the components is below micron (generally < 0.1 μm, for example 100 nm) [24].

2.1 Corrosion Inhibition by Chitosan Composites and Nanocomposites

Chitosan is the second most abundant natural biopolymer after cellulose. It occurs in chitin-rich exoskeletons of marine crustaceans, shrimp and crabs [9, 24]. It is also extracted by N-deacetylation of fungal cell-wall chitin through alkaline treatment and from simple arthropods [24]. Being a polymer, chitosan has (1,4)-β-N-acetyl glycosaminoglycan as the repeating unit (Fig. 1) [25]. Chitosan possesses good antibacterial and antifungal properties hence used widely against skin infections— as drug-carriers in modern therapeutics, in seed treatment and biopesticide to help plants fight against fungal infections [9, 24]. It is utilized as a fining agent in winemaking and also as a preservative. The performance of chitosan as a corrosion inhibitor is a function of the substrate and nature of corrosive environment. In 0.5 M HCl medium, chitosan was found to be very effective in retarding copper corrosion at 25°C [26]. Inhibition efficiency as high as 93.0% was achieved with concentration of 8×10^{-6} M. For steel in strong acid environments, the biopolymer behaves poorly as corrosion inhibitor [27, 28].

The idea of compositing for corrosion protection is basically to improve essential properties like stability, adhesion, barrier, etc. which are essential for effective inhibition. With this in mind, composites and nanocomposites developed with chitosan as the base polymer include poly(*N*-vinyl imidazole) grafted carboxymethyl chitosan composite [29], chitosan/ZnO nanocomposite [22, 30, 31], chitosan-doped-hybrid/TiO$_2$ nanocomposite [32], chitosan/Au nanocomposite [33], chitosan/epoxy composite and nanocomposite [34, 35], poly(aniline-anisidine)/

chitosan/SiO$_2$ composite [36], TiO$_2$-SiO$_2$/chitosan-Lysine composite [37], oleic acid-grafted chitosan/graphene oxide composite [38], and chitosan/silver nanoparticles [39]. The synthesis of polymer composites and nanocomposites intended for use as corrosion mitigating agent is basically done by either *ex situ* method [6, 29], *in situ* method [6, 39], sol gel method [32], or electrochemical deposition method [37]. In the *ex situ* method, the inorganic component is prepared separately and outside the corrosive medium before dispersing into the base matrix [6]. Figure 2 exemplifies the *ex situ* synthesis route for poly(*N*-vinyl imidazole) grafted carboxymethyl chitosan composite. *In situ* method, as the name implies, one of the components is grown directly in the base matrix in the corrosive medium [6].

The *in situ* technique holds the advantage of homogeneous dispersion of components and such dispersion would guard against agglomeration which is a serious challenge in nanocomposite synthesis via *ex situ* technique. In sol-gel technique, solid nanomaterials are dispersed in monomer solution to form a colloidal suspension, i.e., the sol. Thereafter, interconnecting network between phases (gel) and the sol is established by polymerization reactions followed by hydrolysis [32]. This approach had been adopted in the preparation of chitosan-doped-hybrid/TiO$_2$ nanocomposite as illustrated in Fig. 3. Electrochemical deposition is mostly used when the composite or nanocomposite is intended to be used as anticorrosive coatings. The protective film is deposited on substrate to be protected from a solution of ions. There are two ways in which composites and nanocomposites have been applied for metals corrosion protection namely, as inhibitor [23, 29, 39] and as coatings [35, 36, 38]. As inhibitor, Eduok et al. [29] deployed poly (N-vinyl imidazole) grafted carboxymethyl chitosan composite (CMCh-g-PVI) for the protection of API X70 in 1 M HCl. CMCh-g-PVI composite decreased significantly the corrosion of X70 steel compared to equimolar concentration of chitosan and carboxymethyl chitosan in the

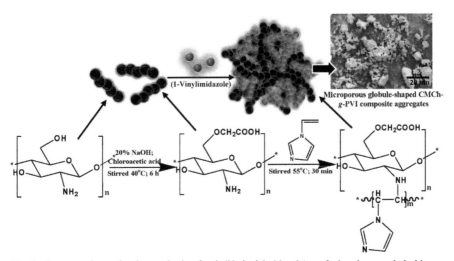

Fig. 2. *Ex situ* pathway for the synthesis of poly(*N*-vinyl imidazole) grafted carboxymethyl chitosan (CMCh-g-PVI) composite. Reproduced with permission from Eduok et al. [29]. Copyright 2018 Elsevier Ltd.

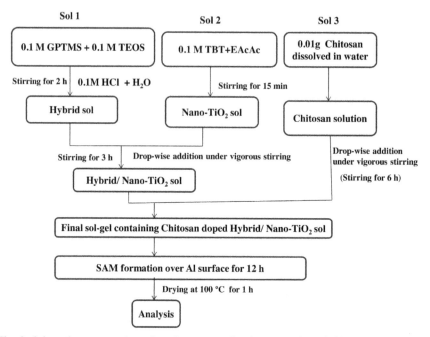

Fig. 3. Schematic representation of coating process for the preparation of chitosan doped-Hybrid/ nano-TiO₂ sol-gel coating on aluminum substrate. 3-glycidoxypropyltrimethoxy silane (GPTMS), tetraethoxysilane (TEOS), titanium (IV) iso-propoxide (TIP), ethyl acetoacetate (EAcAc). Reproduced with permission from Balaji and Sethuraman [32]. Copyright 2017 Elsevier B.V.

same corrosive environment. This was due to the formation of stable CMCh-g-PVI protective polymeric films on the metal surface which was confirmed by means of scanning electron microscopy and atomic force microscopy (Fig. 4(a)).

The adsorbed film raised the charge transfer resistance of the substrate from 89.9 Ω cm² to 1186.0 Ω cm². In a similar approach, Solomon et al. [39] subjected chitosan/ silver nanocomposite (chitosan/AgNPs) synthesized *in situ* using natural honey as the reducing and capping agent to test as an inhibitor for St37 steel in 15% HCl solution using electrochemical impedance spectroscopy (EIS), potentiodynamic polarization (PDP), dynamic electrochemical impedance spectroscopy (DEIS), and weight loss (WL) methods. These testing techniques were supported with surface morphological examination with the help of energy dispersive X-ray spectroscopy (EDS), atomic force microscopy (AFM), and scanning electron microscopy (SEM). Chitosan/ AgNPs showed promising results behaving as effective cathodic type inhibitor particularly at higher temperature and protected the steel surface by formation of a protective film. Just as CMCh-g-PVI, the corrosion resistance property of steel was remarkably enhanced by chitosan/AgNPs composite and the effect of the composite was concentration dependent (Fig. 4(b)).

In most corrosion mitigation practices, polymeric composites and nanocomposites are deployed as coatings and not as corrosion inhibitor [35, 36, 38]. The reason is that polymers generally have the issue of insolubility in aqueous solution. This

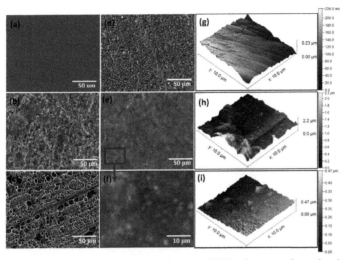

Fig. 4a. SEM micrographs of (a) pre-cleaned and exposed X70 substrate surfaces showing adsorbed poly(N-vinyl imidazole) grafted carboxymethyl chitosan (CMCh-g-PVI) composite film/corrosion product products after 12 h immersion in 1 M HCl (b) without and with (c) chitosan, (d) carboxymethylchitosan, (e) CMCh-g-PVI composite. 3D AFM micrographs (third columns) show topological views of (g) polished/ uncorroded, (h) corroded surfaces in 1 M HCl and (i) protected surface in CMCh-g-PVI polymer-doped solutions. Extracted with permission from Eduok et al. [29]. Copyright 2018 Elsevier Ltd.

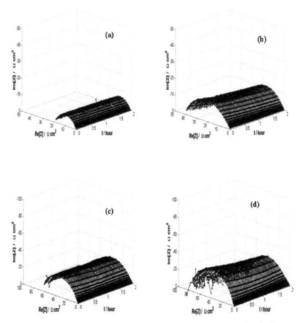

Fig. 4b. DEIS spectra for St37 steel in (a) 15% HCl solution, (b) HCl solution containing 50 ppm AgNPs/chitosan, (c) acid solution containing 500 ppm AgNPs/chitosan, and (d) acid solution containing 1000 ppm AgNPs/chitosan at 25°C after 2 h of measurements. 1000 ppm of the composite exhibited inhibition efficiency of 84.15%. Reproduced with permission from Solomon et al. [39]. Copyright 2016 American Chemical Society.

parameter (solubility) plays a vital role in corrosion inhibition by inhibitors. An inhibitor that is completely soluble in its environment of application would perform better than the one that is partially soluble [40]. Therefore, as coatings, the problem of insolubility is overcome. A nanocomposite consisting of chitosan, oleic acid, and graphene oxide (nano filler) coated on a surface of carbon steel was monitored in 3.5 wt.% NaCl solutions for corrosion resistance. The coated sample maintained long term anticorrosive stability compared to pure chitosan coated specimen [38]. The corrosion resistance of the oleic acid-modified chitosan/graphene oxide coated steel increased by 100 folds relative to chitosan coated sample. Ahmed et al. [33] studied the corrosion resistance strength of NiTi alloy specimens coated with biocomposite of gold nanoparticles and chitosan in Hanks' solution and was found to exhibit better corrosion resistance in different immersion times and pH values. According to John et al. [22] nanostructured chitosan/ZnO nanoparticle films coated on mild steel by sol–gel technique showed excellent chemical stability, oxidation control and enhanced corrosion resistance for the metal substrates in 0.1 N HCl solution. Abd El-Fattah et al. [38] and Ma et al. [35] demonstrated the excellent anticorrosive property of composite and nanocomposite coatings, respectively, formed with chitosan and epoxy coatings on steel in NaCl environment.

In situ chemical oxidative polymerization approach was used by Sambyal et al. [36] to develop a poly(aniline-anisidine)/chitosan/SiO_2 composite in aqueous medium of chitosan and was coated on mild steel and subjected to anticorrosive property test in marine environment. The investigation was for a span of 20 days. The composite exhibited excellent improvement in the corrosion resistant properties of the substrate. Salt spray assessment as per ASTM B117 standards disclosed that the composite coatings can withstand under accelerated corrosion conditions of high salt content and humidity for prolonged periods (Fig. 5). The improved corrosion resistance of the composite coatings was associated with the effective combination of fillers (SiO_2 nanoparticles), and chitosan in poly(aniline-anisidine) matrix. Equally, TiO_2/SiO_2/chitosan-Lysine nanocomposite coated on Ti alloy was found to exhibit high anticorrosion property [37].

Despite the successes reported on biopolymer composites and nanocomposites as metals' protective coatings, there are still some limitations, one of which is the ingression of small corrosive agents like chlorides. It has been found that small anions like chlorides, sulfates, nitrates, perchlorates, etc. can penetrate coatings layer and induce metal surface corrosion, particularly on prolonged immersion [41, 42]. To solve this problem, special dopant anions are incorporated into coatings during synthesis [41, 42]. With such a dopant, it is believed that there would be interchange between the counterion (dopant) in the coatings and small anions present in corrosive solution. In such a case, the ionic active species immobilized in the polymer matrix would force cations instead of the small anions present in the electrolytic solution to diffuse through the coatings [43]. Anions that have been used as dopant include polystyrenesulfonate, phenylphosphinate, sodium dodecyl benzene sulfonate, methane sulfonate, sodium dodecylsulfate, p-toluene-sulfonate, camphosulfonate, polymolybdate, sodium bis(2-ethylhexyl) sulfosuccinate], 5-sulphosalicylate, dioctyl phosphate, poly(methylmethacrylate-co-acrylate), 4-dodecylphenate, and bis(2-

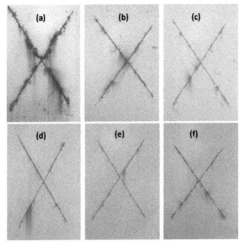

Fig. 5. Photographs of (a) epoxy coated and epoxy with (b) 1.0% (APCS1), (c) 2.0% (APCS2), (d) 3.0% (APCS3), (e) 4.0% (APCS4) and (f) 5.0% (APCS5) loading of poly(anilineanisidine)/chitosan/SiO$_2$ composite coated steel specimens exposed to salt spray fog for 65 days. Extracted with permission from Sambyal et al. [36]. Copyright 2018 Published by Elsevier B.V.

ethylhexyl) sulfosuccinate [41-45]. In the work of Balaji and Sethuraman [32], the corrosion resistance of aluminum coated with undoped hybrid/TiO$_2$ nanocomposite was compared with that coated with chitosan-doped-hybrid/nano-TiO$_2$ composite in 3.5% NaCl medium. It was found that chitosan-doped- hybrid/nano- TiO$_2$ sol-gel coated specimen exhibited better corrosion resistance than the undoped hybrid/ TiO$_2$ nanocomposite coated sample (Fig. 6). Charge transfer resistance of 1.695×10^4 Ω cm^2 was recorded for chitosan-doped-hybrid/nano-TiO$_2$ composite coated sample while that of undoped composite coated specimen was 6875 Ω cm^2.

2.2 Corrosion Inhibition by Carboxymethyl Cellulose Composites and Nanocomposites

Carboxymethyl cellulose (CMC) available as sodium salt share semblance with normal cellulose in terms of structural characteristics but differ in the orientation of the reactive carboxymethyl group, that is, the groups are bound to the hydroxyl functionalities of its cellulosic glucopyranyl moiety (Fig. 1). CMC is the most abundant natural biopolymer and derived from cellulose by alkali-catalyzed reaction with chloroacetic acid [46]. CMC has numerous industrial uses such as viscosity modifier or thickener and stabilizer in different products including ice cream in the food industries [46]. It is a component of products like toothpaste, laxatives, diet pills, water-based paints, detergents, textile sizing, reusable heat packs, and various paper products [46]. In metals corrosion inhibition, it has been examined in acid [47] and alkaline [48] environments but was found to possess moderate inhibiting effect. By incorporation of nanoparticles into CMC matrices, its inhibition efficiency has been significantly boosted. For instance, it was observed [17] that 500 ppm CMC could only suppress the dissolution of mild steel in

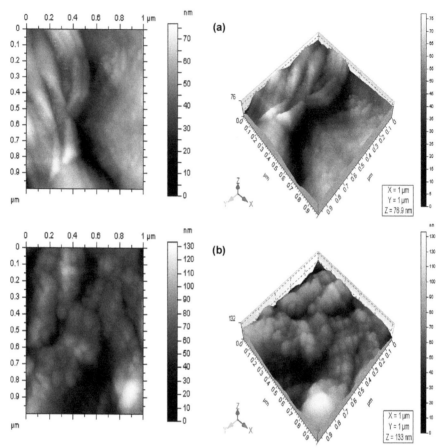

Fig. 6. AFM picture of (a) undoped hybrid/nano-TiO$_2$ coated surface and (b) chitosan doped-Hybrid/nano-TiO$_2$ coated aluminum surface after immersion. Reproduced with permission from Balaji and Sethuraman [32]. Copyright 2017 Elsevier B.V.

2 M H$_2$SO$_4$ solution by 64.8% at 303 K and that the inhibition efficiency decreased with increasing temperature (see Fig. 7). This behavior (decrease in inhibition efficiency with rise in temperature) is consistent with physical adsorption mechanism [47]. In a different investigation [23], the investigators found that by infusing silver nanoparticles into CMC backbone, 500 ppm of the nanocomposite retarded mild steel corrosion in 15% H$_2$SO$_4$ solution (higher acid concentration) by 86.4% at 313 K and the adsorption process was chemisorption, that is, inhibition efficiency increased with rise in temperature (Fig. 7). The primary contributors to these different behaviors of CMC in unmodified and modified forms are the anions in the corrosive medium and the form in which CMC existed in the medium. It has been established that in H$_2$SO$_4$ solution of concentration ≥ 1, metals surfaces acquire net positive charge and are hydrated with sulfate anions [47, 48]. According to Solomon et al. [23, 47], CMC in strong acid solution exists as polycations, meaning CMC polycations were adsorbed on sulfate anions on the mild steel surface which physisorption mechanism

Calculated values of corrosion rates (g/cm²h) and inhibition efficiency (%) for mild steel corrosion in the absence and presence of different concentrations of carboxymethyl cellulose (CMC) at different temperatures from weight loss measurements (Extracted from Solomon et al. [425])

System/concentration	Corrosion rate (g/cm² h)×10⁻³				Inhibition efficiency (%*I*)		
	30 °C	40 °C	50 °C	60 °C	30 °C	40 °C	50 °C
2.0 M H₂SO4	1.79	4.72	15.66	23.60	-	-	-
0.1 g/l CMC	0.86	2.60	8.65	13.12	51.9	44.9	44.8
0.2 g/l CMC	0.84	2.34	7.93	12.30	53.1	50.4	49.4
0.3 g/l CMC	0.75	1.99	6.70	11.92	58.1	57.2	57.8
0.4 g/l CMC	0.71	1.90	6.32	10.95	60.3	59.8	59.6
0.5 g/l CMC	0.63	1.84	6.12	9.26	64.8	61.0	60.9

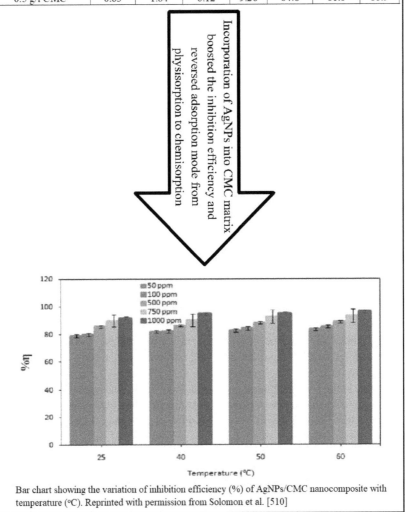

Bar chart showing the variation of inhibition efficiency (%) of AgNPs/CMC nanocomposite with temperature (°C). Reprinted with permission from Solomon et al. [510]

Fig. 7. Influence of AgNPs incorporation on the inhibition efficiency and adsorption mechanism of carboxymethyl cellulose. All the figures were extracted with permission from different Copyright holders.

rightly proposed [47]. Since sulfate anions have lesser shielding power compared to chloride anions [23, 49], the sulfate anions may not appreciably replenish the surface which make it difficult for a sufficient amount of CMC polycations to be adsorbed resulting in the moderate corrosion inhibition. Again, CMC exist in rod-like and most extended form in solutions of low pH and low temperatures but overlap, coil up, and entangle in high pH solutions and at high temperatures to become thermo-reversible gel [23, 47]. This may have been the case in 2 M H_2SO_4 solution and the reason for the reported decline in inhibition efficiency with increase in temperature. Nanoparticles, because of their active properties, can interact chemically with metals surfaces and as a result suppress the opposing force of the positively charged metal surface in the reported system. This will induce the participation of more of the polymers in adsorption process making the nanocomposite a better inhibitor than the unmodified biopolymer. Also, the presence of nanoparticles in CMC backbone could sustain the rod-like and extended form even at elevated temperatures [23] and this prevented the secondary exposure of metal surface to corrosive attack.

2.3 Corrosion Inhibition by Other Biopolymer Composites and Nanocomposites

Other composite and nanocomposite of biopolymers reported as metals corrosion inhibitor are Gum Arabic/silver nanoparticles composite [50], Xanthan gum/poly(acrylamide) [51], and Xanthan gum/polyaniline [52] composites. Gum Arabic (GA) also known as acacia gum, Senegal gum, or Indian gum [53] is a polydispersed exudate (a complex mixture of glycoproteins and polysaccharides) that is obtained from Acacia trees. It contains amino acids bound to short arabinose side chain [50, 54]. Its main field of application is in food industry where its serves as emulsifier and thickener in soft drinks, icing, fillings, soft candy, chewing gum, and other confectioneries [50, 55]. It is also deployed as a binder for watercolor painting due to its high solubility in water [55]. Gum Arabic in polysufone membranes functions as pore-forming and hydrophilic agent [54]. It has been assessed as corrosion inhibitor for steel in HCl [56, 57] and H_2SO_4 [58] media, and for Al in H_2SO_4 medium [58]. Gum Arabic was used to develop a green anticorrosive formulation with silver nanoparticles (AgNPs) and residual natural honey for steel in 15% HCl and 15% H_2SO_4 media [50]. The nanocomposite was very effective in mitigating the steel corrosion in the acid environments, particularly in HCl medium (Fig. 8). It was found that the GA/AgNPs nanocomposite molecules physically adsorbed onto the substrate surface and protected it against corrosion following Langmuir isotherm model. The nanocomposite behaved as a typical mixed type corrosion inhibitor in 15% H_2SO_4 solution but as anodic type in 15% HCl solution. Solomon et al. [50], with the help of X-ray photoelectron spectroscopy, established that both ionic and neutral forms of GA/AgNPS were adsorbed on the steel surface and AgNPs were present on the surface in the form of $Ag°$, Ag_2O, and AgO.

Xanthan gum (XG) is also a polysaccharide discovered by Allene Rosalind Jeanes and brought into commercial production under the trade name Kelzan in the early 1960s [59]. The main chain of XG is the β-(1/4)-D-glycopyranose residues while the side chains are trisaccharide consisting of (3/1)-β-D-mannopyranose,

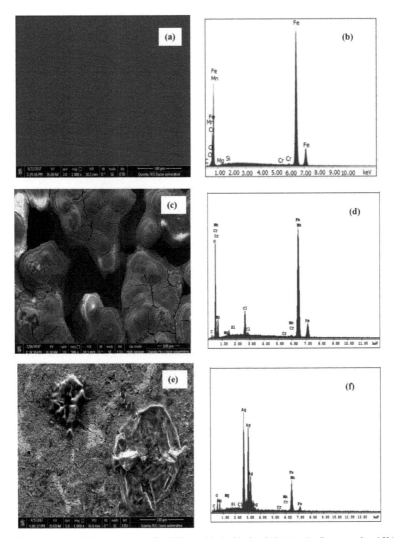

Fig. 8. SEM pictures and EDAX spectra for St37 steel in (a, b) abraded state, (c, d) exposed to 15% HCl solution, and (e, f) exposed to 15% HCl solution containing 1000 ppm of GA-AgNPs for 24 h at 25°C. Extracted from Solomon et al. [50]. Copyright 2017 Elsevier Ltd.

(2/1)-β-D-glucuronic acid, and (4/1)-β-D-mannopyranose [52]. As GA, XG is mostly used as thickening agent and stabilizer in most edible products. It has been reported as metals corrosion inhibitor [60, 61]. Babaladimath et al. [52] prepared polyaniline (PANI) grafted on Xanthan gum backbone as a conducting biopolymer composite and compared the corrosion inhibition property of XG with that of XG-g-PANI in 1 M HCl solution.

Both XG and XG-g-PANI exhibited inhibiting effect but the inhibiting potential of XG-g-PANI significantly surpassed that of XG. From AC impedance measurements, the inhibition efficiency of XG was computed as 65.89% while that of XG-g-PANI was 95.32%. This was similar to the report of Biswas et al. [51] where Xanthan gum grafted poly(acrylamide) was found to perform better as inhibitor for mild steel in 15% HCl than XG alone.

3. Mechanism of Inhibition by Biopolymer Composites and Nanocomposites

Biopolymers after modification to composite or nanocomposite still retain heteroatoms and/or π-electrons and this serves as the adsorption center [23, 24, 40]. As mentioned in the preceding sections, there are two ways in which organic molecules can interact with a metal surface: physical or chemical adsorption [24]. In physical adsorption, inhibitor molecule in charged form is drawn by electrostatic force to a charged metal surface. For chemical adsorption, electron pairs on inhibitor molecule is either transferred or shared with empty d-orbital of metal. The dominant mechanism on a metal surface depends on factors like (i) pH of the aggressive solution, (ii) the true form of the molecule in the aggressive solution, (iii) the anions present in the corrosive environment, (iv) the temperature of the system, and (v) the charge on the metal surface. The charge on a substrate surface can be ascertained by obtaining the difference (d) between the corrosion potential (E_{corr}) and the potential of zero charge (PZC) ($E_{q=0}$) [62]. For $E_{corr} - E_{q=0} = -d$, the metal surface is positively charged while a negatively charged surface will have $E_{corr} - E_{q=0} = +d$ [62]. The form of an organic molecule is dependent on the pH of the solution it is present in. A positively charged surface in a system that has negative anions would unarguably be covered with the anions. If inhibitor species are in protonated form, then physisorption mechanism would be the dominant adsorption mechanism. Neutral inhibitor molecules in a corrosive environment would favor chemisorption mechanism. Nevertheless, the two mechanisms can occur simultaneously on the same metal surface [24]. Both physisorption and chemisorption mechanisms have been observed for biopolymer composites and nanocomposites [21, 23, 29, 39, 50].

4. Conclusions

The chapter examined the application of composites and nanocomposites of biopolymers such as chitosan, carboxyl methylcellulose, gum Arabic, and Xanthan gum for corrosion protection of industrial metal substrates in different corrosive media. From published research data, it can be concluded that compositing presents an efficient way of enhancing the inhibitive capability of biopolymers. By adopting this technique, effective and ecofriendly corrosion inhibitors can be developed for industrial substrates in various corrosive media.

References

[1] Hou, B., X. Li, X. Ma, C. Du, D. Zhang, M. Zheng, W. Xu, D. Lu and F. Ma. 2017. The cost of corrosion in China. npj Materials Degradation 1: 4.

[2] Gharbi, O., S. Thomas, C. Smith and N. Birbilis. 2018. Chromate replacement: what does the future hold? npj Materials Degradation 2: 12.

[3] Singl, W. P. and J. O. Bockris. 1996. Toxicity issues of organic corrosion inhibitors: Application of QSAR model. *In*: Corrosion, Paper No. 225, NACE Houston, TX.

[4] Umoren, S. A. and M. M. Solomon. 2014. Recent developments on the use of polymers as corrosion inhibitors—A review. The Open Materials Science Journal 8: 39–54.

[5] Smith, L. 1999. Control of corrosion in oil and gas production tubing. British Corrosion Journal 34: 247–253.

[6] Umoren, S. A. and M. M. Solomon. 2017. Application of polymer composites and nanocomposites as corrosion inhibitors. pp. 27–58. *In*: Esther Hart (ed.). Corrosion Inhibitors, Principles, Mechanisms, and Applications. New York: Nova Science Publishers Inc.

[7] Tallman, D. E., K. L. Levine, C. Siripirom, V. G. Gelling, G. P. Bierwagen and S. G. Croll. 2008. Nanocomposite of polypyrrole and alumina nanoparticles as a coating filler for the corrosion protection of aluminium alloy 2024-T3. Applied Surface Science 254: 5452–5459.

[8] Elangovan, N., A. Srinivasan, S. Pugalmani, N. Rajendiran and N. Rajendran. 2017. Development of poly(vinylcarbazole)/alumina nanocomposite coatings for corrosion protection of 316L stainless steel in 3.5% NaCl medium. J. Appl. Polym. Sci. 134: 44937.

[9] Xiong, R., A. M. Grant, R. Ma, S. Zhang and V. V. Tsukruk. 2018. Naturally-derived biopolymer nanocomposites: Interfacial design, properties and emerging applications. Materials Science and Engineering R 125: 1–41.

[10] Abdul Khalil, H. P. S., A. H. Bhat and A. F. Ireana Yusra. 2012. Green composites from sustainable cellulose nanofibrils: A review. Carbohydr. Polym. 87: 963–979.

[11] Koeppel, A. and C. Holland. 2017. Progress and trends in artificial silk spinning: a systematic review. ACS Biomater. Sci. Eng. 3: 226–237.

[12] Kuttappan, S., D. Mathew and M. B. Nair. 2016. Biomimetic composite scaffolds containing bioceramics and collagen/gelatin for bone tissue engineering—A mini review. Int. J. Biol. Macromol. 93: 1390–1401.

[13] Umoren, S. A. and M. M. Solomon. 2017. Synergistic corrosion inhibition effect of metal cations and mixtures of organic compounds: A review. Journal of Environmental Chemical Engineering 5: 246–273.

[14] Fares, M. M., A. K. Maayta and J. A. Al-Mustafa. 2012. Corrosion inhibition of Iota-carrageenan natural polymer on aluminum in presence of zwitterion mediator in HCl media. Corros. Sci. 65: 223–230.

[15] Umoren, S. A., O. Ogbobe, I. O. Igwe and E. E. Ebenso. 2008. Inhibition of mild steel corrosion in acidic medium using synthetic and naturally occurring polymers and synergistic halide additives. Corros. Sci. 50: 1998–2006.

[16] Mobin, M. and M. Rizvi. 2016. Inhibitory effect of xanthan gum and synergistic surfactant additives for mild steel corrosion in 1 M HCl. Carbohydr. Polym. 136: 384–393.

[17] Umoren, S. A., M. M. Solomon, I. I. Udosoro, A. P. Udoh. 2010. Synergistic and antagonistic effects between halide ions and carboxymethyl cellulose for the corrosion inhibition of mild steel in sulphuric acid solution. Cellulose 17: 635−648.

[18] Umoren, S. A. 2008. Inhibition of aluminium and mild steel corrosion in acid medium using Gum Arabic. Cellulose 15: 751−761.

[19] Banerjee, S., V. Srivastava and M. M. Singh. 2012. Chemically modified natural polysaccharide as green corrosion inhibitor for mild steel in acidic medium. Corros. Sci. 59: 35–41.

[20] Biswas, A., S. Pal and G. Udayabhanu. 2015. Experimental and theoretical studies of xanthan gum and its graft co-polymer as corrosion inhibitor for mild steel in 15% HCl. Appl. Surf. Sci. 353: 173–183.

[21] Hefni, H. H. H., E. M. Azzam, E. A. Badr, M. Hussein and S. M. Tawfik. 2016. Synthesis, characterization and anticorrosion potentials of chitosan-g-PEG assembled on silver nanoparticles. Int. J. Biol. Macromol. 83: 297–305.

[22] John, S., A. Joseph, A. J. Jose and B. Narayana. 2015. Enhancement of corrosion protection of mild steel by chitosan/ZnO nanoparticle composite membranes. Prog. Org. Coat. 84: 28–34.

[23] Solomon, M. M., H. Gerengi and S. A. Umoren. 2017. Carboxymethyl cellulose/silver nanoparticles composite: Synthesis, characterization and application as a benign corrosion inhibitor for St37 steel in 15% H_2SO_4 medium. ACS Appl. Mater. Interf. 9: 6376–6389.

[24] Umoren, S. A. and U. M. Eduok. 2016. Application of carbohydrate polymers as corrosion inhibitors for metal substrates in different media: A review. Carbohydr. Polym. 140: 314–341.

[25] Ifuku, S. and H. Saimoto. 2012. Chitin nanofibers: preparations, modifications, and applications. Nanoscale 4: 3308–3318.

[26] El-Haddad, M. N. 2013. Chitosan as a green inhibitor for copper corrosion in acidic medium. Int. J. Biol. Macromol. 55: 142–149.

[27] Solomon, M. M., H. Gerengi, T. Kaya, E. Kaya and S. A. Umoren. 2017. Synergistic inhibition of St37 steel corrosion in 15% H_2SO_4 solution by chitosan and iodide ion additives. Cellulose 24: 931–950.

[28] Gupta, N. K., P. G. Joshi, V. Srivastava and M. A. Quraishi. 2018. Chitosan: A macromolecule as green corrosion inhibitor for mild steel in sulfamic acid useful for sugar industry. Int. J. Biol. Macromol. 106: 704–711.

[29] Eduok, U., E. Ohaeri and J. Szpunar. 2018. Electrochemical and surface analyses of X70 steel corrosion in simulated acid pickling medium: Effect of poly(N-vinyl imidazole) grafted carboxymethyl chitosan additive. Electrochim. Acta 278: 302–312.

[30] Al-Naamani, L., S. Dobretsov, J. Dutta and J. G. Burgess. 2017. Chitosan-zinc oxide nanocomposite coatings for the prevention of marine biofouling. Chemosphere 168: 408–417.

[31] Rasool, K., G. K. Nasrallah, N. Younes, R. P. Pandey, P. A. Rasheed and K. A. Mahmoud. 2018. Green ZnO-interlinked chitosan nanoparticles for the efficient inhibition of sulfate-reducing bacteria in inject seawater. ACS Sustainable Chem. Eng. 6: 3896–3906.

[32] Balaji, J. and M. G. Sethuraman. 2017. Chitosan-doped-hybrid/TiO_2 nanocomposite based sol-gel coating for the corrosion resistance of aluminum metal in 3.5% NaCl medium. Int. J. Biol. Macromol. 104: 1730–1739.

[33] Ahmed, R. A., S. A. Fadl-allah, N. El-Bagoury and S. M. F. Gad El-Rab. 2014. Improvement of corrosion resistance and antibacterial effect of NiTi orthopedic materials by chitosan and gold nanoparticles. Appl. Surf. Sci. 292: 390–399.

[34] Abd El-Fattah, M., A. M. El Saeed, A. M. Azzam, A. R. M. Abdul-Raheim and H. H. H. Hefni. 2016. Improvement of corrosion resistance, antimicrobial activity, mechanical and chemical properties of epoxy coating by loading chitosan as a natural renewable resource. Prog. Org. Coat. 101: 288–296.

[35] Ma, I. A. W., Sh. Ammar, K. Ramesh, B. Vengadaesvaran, S. Ramesh and A. K. Arof. 2017. Anticorrosion properties of epoxy-nanochitosan nanocomposite coating. Prog. Org. Coat. 113: 74–81.

[36] Sambyal, P., G. Ruhi, S. K. Dhawan, B. M. S. Bisht and S. P. Gairola. 2018. Enhanced anticorrosive properties of tailored poly(aniline-anisidine)/chitosan/SiO_2 composite for protection of mild steel in aggressive marine conditions. Prog. Org. Coat. 119: 203–213.

[37] Raj, R. M., P. Priya and V. Raj. 2018. Gentamicin-loaded ceramic-biopolymer dual layer coatings on the Ti with improved bioactive and corrosion resistance properties for orthopedic applications. J. Mech. Behavior Biomed. Mater. 82: 299–309.

[38] Fayyad, E. M., K. K. Sadasivuni, D. Ponnamma and M. A. A. Al-Maadeed. 2016. Oleic acid-grafted chitosan/graphene oxide composite coating for corrosion protection of carbon steel. Carbohydr. Polym. 151: 871–878.

[39] Solomon, M. M., H. Gerengi, T. Kaya and S. A. Umoren. 2017. Performance evaluation of a chitosan/silver nanoparticles composite on St37 steel corrosion in a 15% HCl solution. ACS Sustainable Chem. Eng. 5: 809–820.

[40] Umoren, S. A. and M. M. Solomon. 2015. Effect of halide ions on the corrosion inhibition efficiency of different organic species—A review. J. Ind. Eng. Chem. 21: 81–100.

[41] Le, H. N. T., B. Garcia, C. Deslouis and Q. L. Xuan. 2002. Corrosion protection of iron by polystyrenesulfonate-doped polypyrrole films. J. Appl. Electrochem. 32: 105–110.

[42] Rammelt, U., L. M. Duc and W. Plieth. 2005. Improvement of protection performance of polypyrrole by dopant anions. J. Appl. Electrochem. 35: 1225–1230.

[43] Bonastre, J., P. Garcés, F. Huerta, C. Quijada, L. G. Andión and F. Cases. 2006. Electrochemical study of polypyrrole/$PW_{12}O_{40}^{3-}$ coatings on carbon steel electrodes as protection against corrosion in chloride aqueous solutions. Corros. Sci. 48: 1122–1136.

[44] Plesu, N., G. Ilia, A. Pascariu and G. Vlase. 2006. Preparation, degradation of polyaniline doped with organic phosphorus acids and corrosion essays of polyaniline–acrylic blends. Synthetic Metals 156: 230–238.

[45] Shahhosseini, L., M. R. Nateghi, M. Kazemipour and M. B. Zarandi. 2016. Corrosion protective properties of poly(4-(2-Thienyl) benzenamine) coating doped by dodecyl benzene sulphonate. Synthetic Metals 219: 44–51.

[46] Hollabaugh, C. B., H. L. Burt and A. P. Walsh. 1945. Carboxymethylcellulose. Uses and applications. Ind. Engin. Chem. 37: 943–947.

[47] Solomon, M. M., S. A. Umoren, I. I. Udosoro and A. P. Udoh. 2010. Inhibitive and adsorption behaviour of carboxymethyl cellulose on mild steel corrosion in sulphuric acid solution. Corros. Sci. 52: 1317–1325.

[48] Yang, C., Z. Zhang, Z. Tian, Y. Lai, K. Zhang and J. Li. 2018. Influences of carboxymethyl cellulose on two anodized-layer structures of zinc in alkaline solution. J. Alloys Compd. 734: 152–162.

[49] Abd El-Lateef, H. M., A. M. Abu-Dief and B. E. M. El-Gendy. 2015. Investigation of adsorption and inhibition effects of some novel anil compounds towards mild steel in H_2SO_4 solution: Electrochemical and theoretical quantum studies. J. Electroanal. Chem. 758: 135–147.

[50] Solomon, M. M., H. Gerengi, S. A. Umoren, N. B. Essien, U. B. Essien and E. Kaya. 2018. Gum Arabic-silver nanoparticles composite as a green anticorrosive formulation for steel corrosion in strong acid media. Carbohydr. Polym. 181: 43–55.

[51] Biswas, A., S. Pal and G. Udayabhanu. 2015. Experimental and theoretical studies of xanthan gum and its graft co-polymer as corrosion inhibitor for mild steel in 15% HCl. Appl. Surf. Sci. 353: 173–183.

[52] Babaladimath, G., V. Badalamoole and S. T. Nandibewoor. 2018. Electrical conducting Xanthan Gum-graft-polyaniline as corrosion inhibitor for aluminum in hydrochloric acid environment. Mater. Chem. Phys. 205: 171–179.

[53] International Plant Names Index and World Checklist of Selected Plant Families. 2017. http://powo.science.kew.org/taxon/urn:lsid:ipni.org:names:471439-1; Accessed 28 June, 2018.

[54] Manawi, Y., V. Kochkodan, A. W. Mohammad and M. A. Atieh. 2017. Arabic gum as a novel pore-forming and hydrophilic agent in polysulfone membranes. J. Membrane Sci. 529: 95–104.

[55] Smolinske, S. C. 1992. CRC Handbook of Food, Drug, and Cosmetic Excipients. CRC Press, Ohio, p. 7.

[56] Azzaouia, K., E. Mejdoubia, S. Jodehb, A. Lamhamdia, E. Rodriguez-Castellónd, M. Algarrad, A. Zarrouke, A. Errichf, R. Salghig and H. Lgaz. 2017. Eco friendly green inhibitor Gum Arabic (GA) for the corrosion control of mild steel in hydrochloric acid medium. Corros. Sci. 129: 70–81.

[57] Bentrah, H., Y. Rahali and A. Chala. 2014. Gum Arabic as an eco-friendly inhibitor for API 5L X42 pipeline steel in HCl medium. Corros. Sci. 82: 426–431.

[58] Umoren, S. A. 2008. Inhibition of aluminium and mild steel corrosion in acidic medium using Gum Arabic. Cellulose 15: 751–761.

[59] Whistler, R. L. and J. N. BeMiller (eds.). 1993. Industrial Gums: Polysaccharides and their Derivatives. Academic Press, USA, 3rd Edition.

[60] Elkholy, A. E., F. El-Taib Heakal, A. M. Rashad and K. Zakaria. 2018. Monte Carlo simulation for guar and xanthan gums as green scale inhibitors. J. Petrol. Sci. Eng. 166: 263–273.

[61] Mobin, M. and M. Rizvi. 2016. Inhibitory effect of xanthan gum and synergistic surfactant additives for mild steel corrosion in 1 M HCl. Carbohydr. Polym. 136: 384–393.

[62] Gerengi, H., H. I. Ugras, M. M. Solomon, S. A. Umoren, M. Kurtay and N. Atar. 2016. Synergistic corrosion inhibition effect of 1-ethyl-1-methylpyrrolidinium tetrafluoroborate and iodide ions for low carbon steel in HCl solution. J. Adhes. Sci. Technol. 30: 2383–2403.

CHAPTER 3

Development of Green Vapor Phase Corrosion Inhibitors

Victoriya Vorobyova,[1,*] *Olena Chygyrynets*[1] and *Margarita Skiba*[2]

1. Introduction

The atmospheric corrosion of metals is an electrochemical process, which is the sum of individual processes that take place when an electrolyte layer forms on the metal. This electrolyte can be either an extremely thin moisture film (just a few monolayers) or an aqueous film of hundreds of microns in thickness (when the metal is perceptibly wet). Aqueous precipitation (rain, fog, etc.) and humidity condensation due to temperature changes (dew) are the main promoters of metallic corrosion in the atmosphere. Atmospheric corrosion is the most prevalent type of corrosion for common metals because more structures are exposed to air than to any other environment. The costs and tonnage of metal scrap caused by atmospheric corrosion are far higher than any other form of corrosion. Losses due to atmospheric corrosion account for more than half of the total corrosion every year. Carbon steel is the most commonly used metallic material in open air structures and is adopted to fabricate a wide range of equipment and metallic structures because of its low cost and good mechanical strength. Thus, atmospheric corrosion definitely has a tremendous effect on the useful life and durability of structural materials of carbon steel equipment.

As for steel corrosion in solution, it is universally acknowledged that the electrode reactions include anodic and cathodic reactions, which can be reasonably expressed as follows:

$$Fe \rightarrow Fe^{2+} + 2e^-$$ (1)

$$O_2 + 2H_2O + 4e^- \rightarrow 4OH^-$$ (2)

The development of an efficient method of protecting carbon steel from atmospheric corrosion is very important. Several methods of preventing carbon steel

[1] National Technical University of Ukraine "Igor Sikorsky Kyiv Polytechnic Institute", Kyiv, Ukraine. Ave Peremogy 37, Kiev, 03056 Ukraine.
[2] Ukrainian State Chemical-Engineering University, Gagarin Ave. 8, Dnipro, 49005 Ukraine.
* Corresponding author: vorobyovavika1988@gmail.com

from atmospheric corrosion are available. An effective and relatively inexpensive method of controlling corrosion in closed environments is by the use of vapor phase corrosion inhibitors (VCI) [1–8]. There are various kind of evaporating corrosion protection inhibitors had been developed since 1950s in the industrial of metal fields. Commercial VCIs are salts of organic and inorganic acids together with amino compounds, usually supplied as crystals contained in plastic foam supports and applied on carriers such as paper or plastic. VCI are added to closed corrosive environments in relatively small dosages and can reduce significantly the subsequent corrosion rate. VCI are secondary electrolyte layer inhibitors that possess appreciable saturated vapor pressure under atmospheric conditions, thus allowing vapor transport of the inhibitive substance.

The organic bases (usually amines) work as a "passive loader and active in some cases" that carries the organic or inorganic anion when volatilized. The tendency is for the ions to be deposited on the metal surface, forming a uniform and invisible film. Volatile corrosion inhibitors are compounds that have the ability to vaporize and condense on a metallic surface making it less susceptible to corrosion by the formation of a thin protective unimolecular film layer. Organic substances that have been studied as VCI for mild steel are morpholine derivatives and diaminohexane derivatives, fatty acid thiosemicarbazides, cyclohexylamine and dicyclohexylamine, amine carboxylates, ammonium caprylate, benzoic hydrazide derivatives, polyamines, bis-piperidiniummethyl-urea and β–amino alcoholic compounds. Cumulative research over the last 30 years has, unfortunately, has proven that certain specific VPI formulations can, in fact, be toxic. For example, dicyclohexyl ammonium nitrite (DICHAN) has been found to be most effective for inhibiting the atmospheric corrosion of steel, and gained industrial application for several decades. Research confirmed that some N-nitrosoamines, including those generated by DICHAN, were not only carcinogenic, but also hemotoxic as well. Thus, alternative of environmental-friendly VPI is under consideration. Hence, it is important to develop volatile inhibitors of atmospheric corrosion characterized not only by high efficiency and environmental safety but also by a simple procedure of preparation. As promising raw materials for the development of VCI, we can mention organic substances of vegetable origin because, as a rule, natural raw materials contain a large number of biologically active substances and the process of extraction is, as a rule, more cost-efficient than chemical synthesis.

Information of mechanism action corrosion and inhibition processes is very important for proper selection of inhibitors. In general, the inhibition mechanism proposed for VPI is mainly based on its adsorption on the metal surface, whereby a thin monomolecular barrier film should be formed. The metal surface is usually covered with a thin oxide layer under near neutral atmospheric conditions. It is an irregular physical surface with grains of different sizes and orientations, grain boundaries and defects, and most important, a chemistry that is far from that of a pure metal [9]. Thus, the adsorption of VPIs on a metal surface may not be uniform and therefore the protection mechanism of VPIs may involve a combination of passivation and adsorption.

Over the recent years, the use of naturally occurring substances of plant origin, otherwise tagged as "green corrosion inhibitors," has received much more attention because they are incredible sources of natural organic compounds, which are environmentally acceptable, inexpensive, readily available and renewable sources of materials [9–13]. Recently, several researches concern the use of naturally occurring substances as corrosion inhibitors for several metals. Most research is aimed to study the ability to use plants extract as acid corrosion inhibitors or for the corrosion of different metals and alloys in neutral aggressive media. In literature, one can find no information about possible applications of the plants extract and its extracted compounds for the development of volatile corrosion inhibitors. Hence, use of the natural products as eco-friendly and harmless corrosion inhibitors has become popular [4]. Thus, it was shown that the wastes of the production of oil from rapeseeds (*Brassicaceae family*), hop cones (*Humulus lupulus family*), wood bark oil (*Cassia siamea-gonrai, Cassia auriculata, Crataevarelagiosa, Strychonosnuxvomika*) and thyme, the Key Lime Plant (*Citrus latifolia*) can be a source of natural organic compounds for the preparation of VCI [12–21]. The results indicate that the bark oils have significant inhibitive effect. Through these studies, it is agreed that the inhibition performance of plant extract is normally ascribed to the presence in their composition of complex organic species such as aldehydes, terpenoids (thymol, carvacrol), glycosides, nucleosides, ketone, saturated and unsaturated fatty acids. These organic compounds contain polar functions with N, S, O atoms as well as conjugated double bonds or aromatic rings, which are the major adsorption centers. Noticeably, the plant extract is a mixture of various components, which results in the complex inhibitive mechanism. It is rather difficult to determine what components present in plant extract create their relatively high ability to inhibit corrosion.

As one of these sources, we can use products of grapes processing—pomaces (from the *Vitis vinifera* family). Grape pomace is an industrial waste from the wine process and it consists of grape seeds, skin and stems (~ 18–20 kg/100 kg of grapes). The successful utilization of this natural waste may also provide an option for resource recovery. The pomace of grapes, which contain many chemical compounds, can be used as VCI. In the literature, one finds no information about possible applications of the grape pomace and its extracted compounds for the development of VCI. Some works have investigated corrosion inhibition of grape pomace extract on steel in acidic solutions; however, the constituents that provide inhibitive action, the mechanisms and the best condition for inhibition are still unclear.

Although numerous successful reports on plant extracts used as corrosion inhibitors can be found, the phytochemical investigation of the extract is rarely carried out and research is seldom made on the pure compounds present in the plant extract [12–21]. Noticeably, the plant extract is a mixture of various components, which results in the complex inhibitive mechanism. It is rather difficult to determine what components present in plant extract create their relatively high ability to inhibit corrosion. A better way is to isolate the components and investigate the inhibition of each single component, but it is still difficult to isolate all the components. Thus, testing the inhibition potential of major components using available pure compounds could be an alternative choice to explore the corrosion inhibition of plant extract.

Thus, evaluating the corrosion inhibition effect of grape pomace extract (GPE) as a green vapor phase corrosion inhibitor of mild steel and investigating the constituents that provide its inhibitive action is an important topic of research. For this purpose, the compounds were identified and quantified by GC-MS analysis. The corrosion inhibition efficiency of the grape pomace extract and of its main compounds was characterized by using electrochemical methods and was compared to their inhibition activity evaluated by the method of accelerated tests under simulated operational conditions. The electro-chemical behaviour analysis and microscopic surface observation after immersion in the corrosive medium were combined to discuss the role of the main molecules in the corrosion protection brought by the extract.

2. Volatile Components of Grape Pomace Extract

The organic solvent 2-propanol was applied for soxhlet extraction. Grape pomace samples (5.0 g) were placed inside a thimble made by thick filter paper, loaded into the main chamber of the soxhlet extractor, which consisted of an extracting tube, a glass balloon and a condenser. The total extracting time was 6 h, with the solvent continuously refluxing over the sample (grape pomace).

The composition of volatile substances of the extract was studied by means of chromatography and mass spectrometry methods, using a FINIGAN FOCUS gas chromatograph. About 10 μl of the grape pomace extract sonicated with n-hexane were analyzed by GC–MS using Shimadzu Model GC-17A equipped with flame ionization detection (FID) and a CBP-5 capillary fused silica column (25 m, 0.25 mm i.d., 0.22 μm film thickness). The oven temperature was held at 50°C for 2 min then programmed at 10°C/min to 250°C, and then held for 20 min. Other operating conditions were as follows: carrier gas He (99.99%), inlet pressure 76 kPa, with a linear velocity of 20 cm/s; injector temperature 250°C; detector temperature 310°C; split ratio 1:25.

The components were identified by comparing the peak retention times in the chromatogram and the complete mass-spectra of individual components with the corresponding results for pure compounds in the NIST-5 Mass Spectral Library. In the volatile fraction of the grape pomace extract, 22 compounds were detected by the chromatography and mass spectrometry method (Fig. 1, Table 1). This result indicates that grape pomace extracts contain different classes of organic substances in their composition, which can act as corrosion inhibitors.

According to the chromatogram, nine aldehydes are identified by GC-MS in the pomaces grape extract namely, Hexanal (9.1%), Heptanal (8.7%), 2-Phenylacetaldehyde (6.5%), Butanal (4.8%), Nonanal (7.0%), (2E)-3,7-Dimethyl-2,6-octadien-1-ol (2.3), 4-Hydroxy-3-methoxybenzaldehyde, terpenes and saturated and unsaturated fat acids (34%) represented by hexadecanoic acid (5.4%), (9Z)-Octadec-9-enoic acid (6.1%), (9Z,12Z)-9,12-Octadecadienoic acid (4.2%). Several peaks that were present in trace amounts (< 0.01%) remained unidentified (i.e., they did not match any hit in the library search) and their elucidation required the use of more advanced techniques.

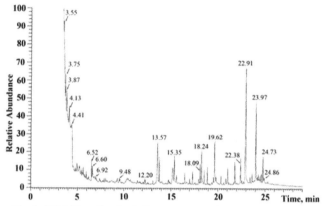

Fig. 1. GC–MS profile of volatile components of grape pomace extract.

Table 1. GC-MS analysis result of GPE.

Name of compound	Holding time t, min	Weight ratio, %
Hexanal	3.55	9.1
Heptanal	3.75	8.7
2-Phenylacetaldehyde	3.87	6.5
Butanal (Butyraldehyde)	4.13	4.8
Nonanal	4.41	7.0
(2E)-3,7-Dimethyl-2,6-octadien-1-ol	6.52	2.3
Propane-1,2,3-triol (Glycerol)	6.60	0.9
4-Hydroxy-3,5-dimethoxybenzaldehyde (Syringaldehyde)	6.92	1.1
(2E)-3-Phenylprop-2-enal (Cinnamaldehyde)	9.48	0.9
4-Hydroxy-3-methoxybenzaldehyde (Vanillin)	12.20	1.1
2-methyl-2-propanyl 4-hydroxy-4-methyl-5-hexanoate	13.57	4.8
Heptacosane	17.31	3.2
(9Z,12Z)-9,12-Octadecadienoic acid (Linoleic acid)	18.09	4.2
Hexadecanoic acid (Palmitic acid)	18.24	5.4
(9Z)-Octadec-9-enoic acid (Oleic acid)	19.62	6.1
(1R,4E,9S)-4,11,11-Trimethyl-8-methylidenebicyclo[7.2.0]undec-4-ene (Caryophyllene)	21.01	1.7
(Naphthalene-1-yl) (1-pentyl-1H-indol-3-yl) methanone	22.86	6.2
2-(4-Methyl-1-cyclohex-3-enyl)propan-2-ol (α-Terpineol)	23.94	9.5
1-Methyl-4-(1-methylethylidene)cyclohexan-1-ol (γ-Terpineol)	24.01	4.2
(1R,3aR,5aR,5bR,7aR,9S,11aR,11bR,13aR,13bR)-3a,5a,5b,8,8,11a-hexamethyl-1-prop-1-en-2-yl-1,2,3,4,5,6,7,7a,9,10,11,11b,12,13,13a,13b-hexadecahydrocyclopenta[a]chrysen-9-ol (Lupeol)	24.52	9.5
Lup-20(29)-ene-3β,28-diol (Betulin)	27.41	3.8

The results of the weight ratio are given in Table 1 and show that among the identified aldehydes, the major compounds are Hexanal, Heptanal, 2-Phenylacetaldehyde, Butanal, Nonanal, (2E)-3,7-Dimethyl-2,6-octadien-1-ol, (2E)-3-Phenylprop-2-enal (Cinnamaldehyde), 4-Hydroxy-3-methoxybenzaldehyde (Vanillin). The aim of this section is to compare the inhibition provided by grape pomace extract constituents to find the main constituent responsible for corrosion inhibition properties of the plant extract. Therefore, for inhibition characterization the main compounds, namely, 2-phenylacetaldehyde and hexanal, were chosen as representative molecules of the grape pomace extract. The chemical formulas of the chosen inhibiting molecules are given in Table 2.

Table 2. Chemical formulae of the main compounds of GPE.

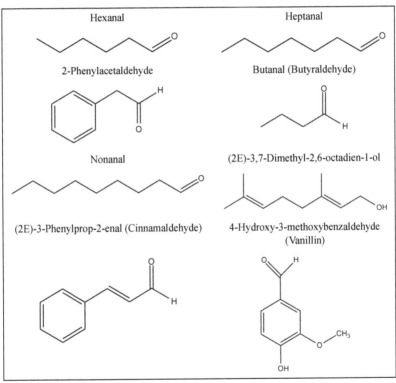

3. Effect of Grape Pomaces Extract and its Main Compounds on Corrosion Rate and Inhibition Efficiency

3.1 Gravimetric Measurements

Inhibitor effectiveness of the grape pomace extract as a vapor phase inhibitor of atmospheric corrosion of steel was evaluated with the method of accelerated tests under condition of condensation of moisture. The final geometrical area was 25 cm². The gravimetric measurement was conducted by suspending the samples in a

250 cm³ conical flask with a tight-fitting rubber cork containing a small dish. The VCI were placed in the dish (2 ml) for a certain time (depending on the experiment from 24 to 72 hours) to form a protective film. After inhibitor film-forming period, 15 cm³ of deionized water were added. The test process included cyclic warming and cooling of the samples in a corrosion testing chamber with varying humidity. One cycle included an 8 h exposure in the thermostat (40 ± 1°C), and a 16 h exposure at room temperature (25°C). The total duration of the tests was 21 days. To prove reproducibility, gravimetric measurements were repeated twice.

Figure 2 represents inhibition efficiency values obtained from the weight loss in conditions of periodic condensation of moisture after period of film-forming in the presence of different concentrations of grape pomace extract and major components. Clearly, inhibition efficiency increases with an increase of the inhibitor concentration, i.e. the corrosion inhibition enhances with the inhibitor concentration. This behavior is explained with the fact that the adsorption amount and coverage of inhibitor on steel surface increases with the inhibitor concentration.

At 80 mg L⁻¹, the maximum *IE* is 92.08% for grape pomace extract; 72.24% for 2-Phenylacetaldehyde; 44.6% for Hexanal, which indicates all compounds act as moderate corrosion inhibitors for steel. Figure 2 also shows that the values of inhibition efficiency follow the order: grapes pomace extract > 2-phenylacetaldehyde > hexanal. This result implies that the grape pomace extract exhibits better inhibition performance than its major components. Similar comparative results between plant extract and its major components have also been reported for the *coffee extract* [27] and *Artemisia pallens* [26].

Influence of the pre-treatment immersion time of steel in the volatile phase of the GPE on corrosion effect under the conditions of periodic moisture condensation was investigated. It should be noted that the corrosion rate decreases with increasing the time film-formation. The inhibition efficiency (IE) increases in the range 65.32– 96.39%. The inhibition efficiency after 72 h of GPE film-forming was higher

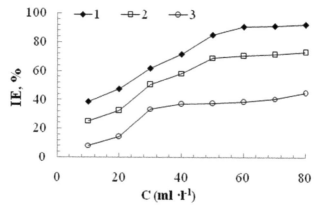

Fig. 2. Relationship between inhibition efficiency (gw) and concentration of inhibitor (c) (1-Grape pomace extract, 2-2-phenylacetaldehyde, 3-hexanal) in the conditions of periodic condensation of moisture for 21 days (weight loss method, pre-treatment immersion time 48 h).

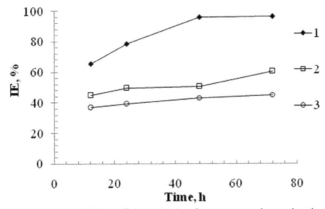

Fig. 3. Relationship between inhibition efficiency (gw) and pre-treatment immersion time St3 in vapor
phase of inhibitor (1-Grape pomace extract, 2-2-phenylacetaldehyde, 3-hexanal).

than that after 12 and 46 h of GPE film-forming (Fig. 3). This suggested that the
corrosion protectiveness of GPE film-formed on the steel surface was enhanced by
prolonging the GPE treatment. The protective film on the steel surface is formed
after 40–48 h of its treatment with volatile compounds of the extract. The effect
of immersion time in the vapor phase of extract on the weight loss rate indicated
that GPE does not only keep its inhibitive activity for steel, but it also improves
its effectiveness over the long-term immersion due to synergistic influence of the
main compounds which offer an additional protection. It should be noted that the
treatment of specimens in the vapor phase of individual 2-propanol does not improve
the corrosion resistance of the metal. This enables us to recommend this extract for
the corrosion protection of mild steel.

An important characteristic of the films formed on the metal surfaces in the
vapor of the volatile inhibitor is their ability to preserve protective properties (after
effect) in the course of time in the absence of VIC in corrosive media. Therefore, we
investigated the aftereffect of films formed on the metal surface after preliminary
holding of the specimens in the vapor of the 2-propanol grape pomace extract. It was
shown (Table 3) that these films give the required aftereffect under the conditions of
periodic condensation of moisture over 1 M Na_2SO_4 aqueous solutions.

Table 3. After effect of the films formed on st3 steel from the vapor phase of the 2-propanol GPE (Time
of formation 120 h; Time of experiment 504 h).

Testing conditions			Corrosion rate K_m^- g/(m²·h)	Inhibition efficiency/%
Periodic moisture condensation	3% NaCl	In the vapor phase of 2-propanol GPE	0.0356	90.3
		Without inhibitors	0.3663	–
	1 M Na_2SO_4	In the vapor phase of 2-propanol GPE	0.0188	91.8
		Without inhibitors	0.2346	–

The protective chemisorption films formed on the surface serve as a barrier for the penetration of electrolytes into the metal. The presence of the after effect shows that the nature of the inhibiting action is not connected to the electrostatic adsorption. It is rather caused by the formation of a chemisorptions layer of organic compound on the metal surface. These results infer that the inhibition of corrosion steel is due to blocking of steel surfaces by chemically adsorbed molecules.

3.2 Electrochemical Measurements

An electrochemical measurement was conducted in simulated atmospheric corrosion solution. The stimulated atmospheric corrosion solution was prepared by using double-distilled water containing 7.1 g Na_2SO_4 mg/L ($\sim$ 0.5 N Na_2SO_4). Electrochemical experiments were carried out in a conventional three-electrode cell with a platinum counter electrode (CE) and a saturated calomel electrode (SCE) coupled to a fine Luggin capillary as the reference electrode. To minimize the ohmic contribution, the tip of Luggin capillary was kept close to the working electrode (WE). The carbon steel working electrode was designed with a fixed exposed surface area of 0.385 cm^2. As a specific feature of our electrochemical investigations, we noted that the disk-shaped surface of the end face of the working electrode was immersed in surface layers of the working solution by at most 1–2 mm. Polarization measurements were performed to monitor the mechanism of anodic and cathodic partial reactions as well as identifying the effect of an inhibitor on each partial reaction (Fig. 4–6).

It was found that molecules of the inhibitors were able to induce some corrosion activity during the first stage of contact with the brass surface. From the potentiodynamic polarization curves, it can be seen that the extract and basic compounds caused a decrease in both anodic and cathodic current densities, most

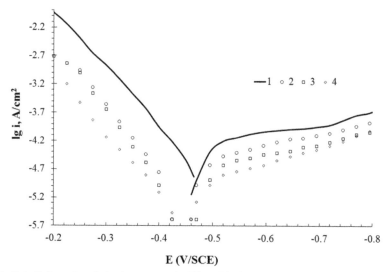

Fig. 4. Potentiodynamic polarization curves of mild steel in 0.5 M Na_2SO_4 without (1) and with film formed after 24 (2), 48 (3) and 72 (4) hours in the vapor phase of GPE.

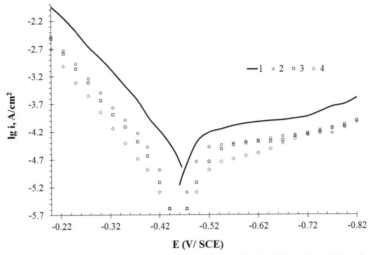

Fig. 5. Potentiodynamic polarization curves of mild steel in 0.5 M Na_2SO_4 without (1) and with film formed after 24 (2), 48 (3) and 72 (4) hours in the vapor phase of 2-phenylacetaldehyde.

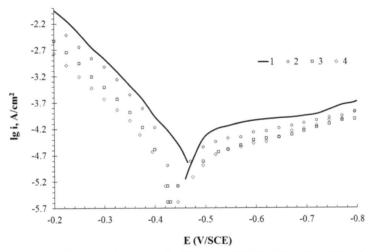

Fig. 6. Potentiodynamic polarization curves of mild steel in 0.5 M Na_2SO_4 without (1) and with film formed after 24 (2), 48 (3) and 72 (4) hours in the vapor phase of hexanal.

likely due to the adsorption of the organic compounds present in the extracts at the active sites of the electrode surface. Both the cathodic slopes and the anodic slopes do not change, which indicates that the mechanism of the corrosion reaction does not change, and the corrosion reaction is inhibited by a simple adsorption mode. The presence of protective film on the surface that is formed in vapor phase of extract results in marked shift in the cathodic branches and to a lesser extent in the anodic branches of the polarization curves. Moreover, in the presence of grape pomace extract the values of corrosion potential E_{corr} are nearly constant; therefore, the GPE could be classified as a mixed-type inhibitor with predominant cathodic effectiveness.

The inhibitive action for these constituents increases in the following order: GPE > 2-phenylacetaldehyde > hexanal. This order reflects the important role played by the 2-phenylacetaldehyde molecules. It has the highest inhibition efficiency among other constituents. The electrochemical impedance spectroscopy was recorded at open circuit potential with a 10 mV AC perturbation at frequency ranging from 100 kHz to 0.05 mHz with 10 points per decade. The test device and the cell configuration for the EIS measurements were the same as those for the polarization curve tests. To prove reproducibility, the polarization curve tests and the EIS measurements were repeated three times.

Figure 7 represents the Nyquist diagrams for steel sample in 1 N Na_2SO_4 without protective film (curve 1) and with a film formed after holding for 48 h in the vapor phase of GPE (curve 2). The most common equivalent circuit used to model corrosion of bare metal in aqueous electrolyte is the *Randles* circuit, shown in Fig. 8.

The Nyquist diagram of steel in the absence of protective film describes a capacitive arc with high frequency values, which is followed by a straight line at lower frequency values (Table 4). In the presence of protective film, the impedance spectrum switches to double-capacitive semicircles. The disappearance of the diffusion features in the presence of protective film indicates that the corrosion

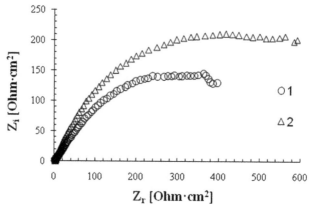

Fig. 7. Nyquist plots measured at E_{corr} for St3 steel in 1 N Na_2SO_4 without protective film (curve 1) and with a film formed after holding for 48 h in the vapor phase of the volatile compounds of GPE.

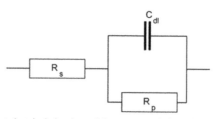

Fig. 8. Equivalent electrical circuit used for computer fitting of the experimental data.

Table 4. Typical parameters obtained from EIS fitting results of St3 steel in 1 N Na$_2$SO$_4$ without protective film and with a film formed after holding for 48 h in the vapor phase of GPE.

Sample	EIS fitting parameters			
	R$_1$, Ohm.cm^{-2}	R$_2$, Ohm.cm^{-2}	CPE$_1$, Ohm^{-1} s. cm^{-2}	W*, Ohm.cm^{-2}
Without protective films	3.76	586	26	-
With protective films	3.97	812	20	-

*W—resistance of Warburg, which is connected with diffusion limitation (resistance of electrolyte in pores). The results obtained from both weight loss and potentiodynamic polarization tests are in good agreement, and the abstracts inhibition action could also be evidenced by surface SEM and AFM images.

reaction is inhibited by protective film, which makes the diffusion process no longer the controlling process.

3.3 Scanning Electron Microscope (SEM) Surface Examination

The surface morphology and the coating were examined by FEI E-SEM XL 30. For SEM images, 1 cm^2 samples were imaged. SEM analyses were conducted in order to characterize the protective layer that is formed on the low carbon steel surface. SEM images of initial surface and surface after 48 hours of film-forming are shown in Fig. 9–11. The surface morphology of the sample before exposure to volatiles of extract and two compounds indicates that there are a few scratches from the mechanical polishing treatment (Fig. 9a). Figure 9a displays a freshly polished steel surface. The following images (Fig. 9b) are of the steel surface after 48 h of exposure for the film-forming of the grape's pomace extract. It shows a thin film which covers the surface. As shown in Fig. 9b, the steel sample appeared to be smooth and without any visible traces of corrosion products, due to the formation of a protective inhibition layer on the metallic surface, which retarded the dissolution process. Presumably, the inhibitory effect was performed via the adsorption of compounds present in the grape pomace extract onto the steel surface. The steel surface after 48 h exposition for the

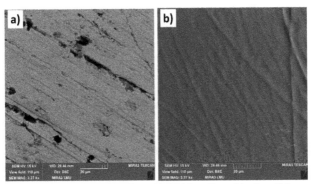

Fig. 9. SEM images of the carbon steel surface: (a) initial surface; (b) after 48 h exposure for GPE film-forming.

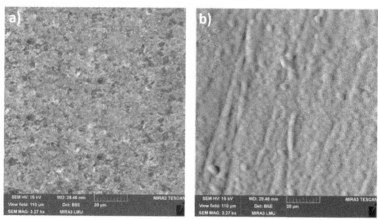

Fig. 10. SEM images of the carbon steel surface after 24 h (a) and (b) 48 h exposure for 2-phenylacetaldehyde film-forming.

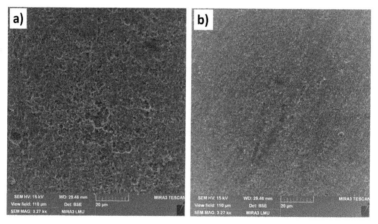

Fig. 11. SEM images of the carbon steel surface after 24 h (a) and (b) 48 h exposure for hexanal film-forming.

film-forming in the vapor phase of 2-phenylacetaldehyde and grape pomace extract had the typical structures (Fig. 10a, b). A totally different situation was observed after 48 h of film-forming in the vapor phase of hexanal (Fig. 11). The steel surface after 48 h of film-forming in vapor phase of 2-phenylacetaldehyde was slightly rougher than that after forming film in volatile compounds of hexanal. The particles are less distributed and rather uniform, and the outer surface of the particles is enwrapped tightly and regularly.

3.4 Atomic Force Microscope (AFM) Surface Examination

Surface analysis was carried out by AFM technique in order to evaluate the surface morphology of the steel samples after 48 h of film-forming in the vapor phase of inhibitors. Samples of the dimension 1.0 cm × 1.0 cm × 0.06 cm were abraded with emery paper (grade 320-500-800) and then washed with distilled water and acetone. After immersion in the vapor phase of inhibitors, the specimens were analyzed using SPA-400 SPM Unit atomic force microscope (AFM).

The atomic force microscope provides a powerful means of characterizing the microstructure. The two-dimensional AFM image of steel surface before exposure to volatiles of extract was shown in Fig. 12a. Figure 12b shows the spherical or bread-like particles that appear on the surface, which do not exist in Fig. 12a. Therefore, it may be concluded that these particles are the adsorption film of the inhibitor, which efficiently inhibits the corrosion of steel. Figure 12b also shows that a uniform film was formed. The inhibition efficiency of grapes pomace extract is attributed to joint adsorption of some of its phytochemical constituents. Therefore, the use of 2-propanol GPE as an inhibitor for the corrosion of mild steel from corrosion under the conditions of periodic moisture condensation is recommended.

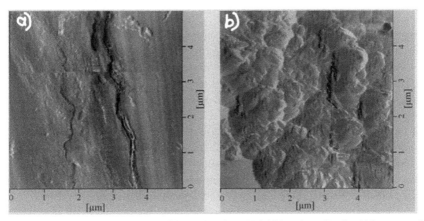

Fig. 12. AFM two-dimensional images of the steel surface: (a) initial surface; (b) after 48 h exposure for GPE film-forming.

3.5 FT-IR Analysis

FT-IR analyses was carried out by Bruker Tensor 27 FT-IR spectrometer in attenuated total reflection mode (Pike Technologies, GladiATR for FTIR with diamond crystal) and using spectral range of 4000–400 cm^{-1} with a resolution of 4 cm^{-1}.

Fourier transformation infrared spectroscopy was used to identify the functional groups of components which are present in the grape pomace extract and those were adsorbed onto surface of steel. FT-IR analysis of the carbon steel specimens in the absence and after the formation of a film for 72 h in the presence of extract was carried out between 600 and 4000 cm^{-1}. The presence of the C-H, O-H, C=O, C=C and CH$_2$ groups indicates that 2-propanol rape extract contains compounds which, according to the results of GC-MS, have been identified as major constituents.

In the FT-IR spectrum of GPE, presented in Fig. 13, several characteristic absorption bands were identified, i.e., stretching vibrations of intermolecular associated hydroxyl groups, v (O–H) at 3368 cm^{-1}, symmetrical and asymmetrical

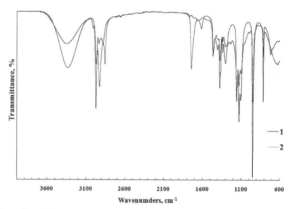

Fig. 13. FT-IR absorption of 2-propanol GPE (1) and spectra displayed on the surface of Steel 3, treated in volatile compounds of the grapes' pomaces extract (2).

vibrations for alkyl units v (CH$_2$) at 2926 cm^{-1} and 2851 cm^{-1}, vibrations for aromatic rings v (C=C) at 1514 cm^{-1} and phenols moiety v (C–OH) at 1147 cm^{-1}. The band at 1640 cm^{-1} is assigned to conjugated C=O stretching vibration. The absorption bands within the interval from 1750–1600 cm^{-1} are related to the stretching vibrations of carbonyl groups of aldehydes. The band at 1452 cm^{-1} can be attributed to C=C-C aromatic ring stretching. The obtained results indicate that after 72 hours (Fig. 13), the formation of the film lead to a decrease in the intensity of the oscillation in the region of 1700–1580 cm^{-1}. The new band at 1690 cm^{-1} is characteristic of v (C = C) groups which maybe corresponding to new groups of compounds. This indicates the possible chemical change of the main compounds of GPE once adsorbed on the mild steel surface and explains the highest protection values that are achieved only in 48 h.

4. Quantum Chemical Calculations

The quantum chemical calculations serve as a theoretical model for establishing a correlation between molecular structure and corrosion inhibition efficiency. Quantum chemical calculations have been performed by the HyperChem 7 package. The geometry optimization was obtained by application of the restricted Hartree-Fock method (RHF) using MNDO approach with PM3 parameterization.

Corrosion inhibition effectiveness of molecules can be compared with the help of various electronic structure parameters. The most popular parameters among these are the eigenvalues of highest-occupied (HOMO) and lowest-unoccupied (LUMO) molecular orbitals, HOMO–LUMO gap, electronegativity and chemical hardness [30]. According to *Koopman's* theorem, The frontier orbital energies E_{HOMO} and E_{LUMO} are related to the ionization potential, I, and the electron affinity, A, of iron and the inhibitor molecule by the following relations [31]: $A = -E_{LUMO}$, $I = -E_{HOMO}$. *Pearson* and *Parr* presented operational and approximate definitions using the finite differences method depending on electron affinity (A) and ionization energy (I) of any chemical species (atom, ion or molecule) for chemical hardness (η), and chemical potential (χ) [25]

$$\chi = -\mu = \frac{1}{2}(I + A) \tag{3}$$

The hardness η of an electronic system is defined as [30]:

$$\eta = \frac{1}{2}(I - A), \tag{4}$$

The global softness (S) is the inverse of global hardness and is given as follows:

$$S = \frac{1}{2\eta} \tag{5}$$

It has been reported that the most stable molecular structure has the largest HOMO-LUMO energy gap [29]. Therefore, an electronic system with a larger HOMO-LUMO gap should be less reactive than one having a smaller gap. This relationship is based on the Maximum Hardness Principle, which states that there seems to be a rule of nature that molecules arrange themselves so as to be as hard as possible.

This principle is among the most widely accepted electronic principles of chemical reactivity and a formal proof of this principle was reported [31]. If bulk iron metal and the inhibitor molecule are brought together, the flow of electrons will occur from the molecule of lower electronegativity to the iron that has higher electronegativity until the value of the chemical potential becomes equal [28, 29]. The global electrophilicity index (ω) is given by:

$$\omega = \frac{(I + A)^2}{8(I + A)} \tag{6}$$

Nucleophilicity *(ε)* is physically the inverse of electrophilicity (*1/ω*). The electron charge transfer, ΔN, from base B to acid A, and the associated energy change ΔE is given as follows:

$$\Delta N = \frac{\mu_B - \mu_A}{2(\eta_A + \eta_B)} \tag{7}$$

$$\Delta E = \frac{(\mu_B - \mu_A)^2}{2(\eta_A - \eta_B)} \tag{8}$$

A, B—Indices in formulas (7 and 8) are:

A—characteristics of the molecule of the test substance;

B—characteristics of the elemental lattice of the surface of iron.

Substituting the subscripts A and B by *mol* and *met* to designate a metal surface and a molecule, respectively, and replacing the μ by χ gives:

$$\Delta N = \frac{\chi_{met} - \chi_{mol}}{2(\eta_{met} + \eta_{mol})} = \frac{\Phi - \chi_{mol}}{2\eta_{mol}} \tag{9}$$

$$\Delta E = \frac{(\chi_{met} - \chi_{mol})^2}{2(\eta_{met} - \eta_{mol})} = \frac{(\Phi - \chi_{mol})^2}{4\eta_{mol}} \tag{10}$$

In the second equality of equations 9 and 10, the electronegativity of metal surface is replaced by the work function for Fe (110) surface, Φ, theoretically equals 4.82 eV and the hardness, η_{metal} which equals 0 eV for bulk metals [25].

The parameters which were defined above are the most important parameters considered in scientific researches about corrosion inhibition. It should be clear that these parameters provide information about electron donating ability of chemical species and the most notable feature of a corrosion inhibitor is its electron donating ability to metals because electron donor chemical species prevent the oxidation of metals.

In recent years, theoretical quantum chemical calculation was used to predict the adsorption center and explain the mechanism of corrosion inhibition effects. It is a useful method to investigate the mechanisms of reaction by calculating the structure and electronic parameters, which can be obtained by means of theoretical quantum theory [24–28]. In chemical model of adsorption, a chemical bond is presumed to be formed as a result of the donation of electrons from the filled HOMO of the inhibitor molecules to the vacant d-orbitals of the metal. It is also possible that back bonding can result through electron donation by the filled metal d-orbitals to the vacant LUMO of the inhibitor.

Thus, the adsorption power of hexanal and 2-Phenylacetaldehyde using quantum chemical concepts and the density functional theory was applied to study the geometrical and electronic structures of the compounds. The electric/orbital density distributions of HOMO and LUMO for hexanal and 2-Phenylacetaldehydemolecules are shown in Fig. 14. It is found that the electron density of the frontier orbital is well proportioned. That is, there is electron transferring in the interaction between the inhibitor molecule and metal surface. The values of energy of highest occupied molecular orbital (E_{HOMO}), energy of lowest unoccupied molecular orbital (E_{LUMO}) and the separation energy ($\Delta\varepsilon = E_{LUMO} - E_{HOMO}$) are also presented in Table 5.

From Table 5, E_{HOMO} shows almost no difference between hexanal and 2-phenylacetaldehyde. On the other hand, E_{LUMO} obeys the order: hexanal <

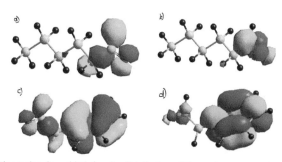

Fig. 14. The frontier molecular orbital density distribution of the main components of GPE using DFT: HOMO and LUMO of hexanal (a, b) and 2-phenylacetaldehyde (c, d).

Table 5. Calculated quantum chemical properties for the most stable conformations of the major effective components of the grape pomace extract.

Molecule	E_{HOMO} (eV)	E_{LUMO} (eV)	$\Delta\varepsilon_{H-L}$ (eV)
Hexanal	−10.771	0.864	−11.63
2-Phenylacetaldehyde	−9.908	0.125	−10.03

2-phenylacetaldehyde, which is in complete agreement with the inhibition efficiency order of 2-phenylacetaldehyde > hexanal.

The energy gap $\Delta\varepsilon_{H-L}$ (eV) value was also related to the inhibition efficiency. A low value of $\Delta\varepsilon$ leads to higher inhibition efficiency because less energy will be required to remove the electron from last occupied orbital. The values in Table 5 show the following increasing reactivity abilities of the molecules towards steel: 2-Phenylacetaldehyde > Hexane.

Corrosion inhibition of mild steel by organic inhibitors (present in plant extract) is a complex process and is mainly influenced by charge transport. Values of quantum chemical indices such as electronegativity (χ), hardness (η), electrophilicity index (ω), electron charge transfer (ΔN) and energy change (ΔE) associated with the electron charge transfer from molecule to steel surface are given in Table 6.

According to some studies, the parameter of μ is related to the chemical potential, and higher value of μ means better inhibitive performance. Electrophilicity (ω) is an index which measures the propensity of chemical species to accept electrons. Thus, a good nucleophile is characterized by low values of μ and ω, whereas a good electrophile is characterized by high values of μ and ω. It is clear in Table 6 that the molecules have low electrophilicity index values and are good nucleophiles. Values of ΔN exhibit inhibitive performance resulting from electron donations. If $\Delta N < 3.6$, the inhibition efficiency increases with the increase in electron-donation ability to the metal surface. It can be seen from Table 6 that the values of ΔN between 2-phenylacetaldehyde and Hexanal have a difference.

In the present study, inhibition efficiency follows the order: 2-phenylacetaldehyde > hexanal. Thus, there is a good correlation between inhibition efficiency and the parameters of E_{LUMO}, $\Delta\varepsilon_{H-L}$, ΔN and ΔE. As shown by the quantum chemical calculations, the inhibition efficiency of the main compounds identified in the GPE depends on the molecule structure. The molecules studied possess charge transfer abilities towards mild steel. Therefore, inhibitor molecules of GPE are the electron donors and the mild steel surface is an acceptor. Thereby, binding the inhibitor molecules to the mild steel results in protective adsorption layer against corrosion. In addition, the theoretical inference is in good correspondence with experimental data.

Table 6. Quantum chemical parameters for inhibitor molecules and adsorption energies between the molecules and Fe (001) plane.

Molecule	μ (eV)	η (eV)	ω	ΔN	ΔE
Hexanal	−4.95	5.81	1.238	0.175	0.00077
2-Phenylacetaldehyde	−4.89	5.01	1.222	0.210	0.00025

5. Explanation of Inhibition

Grape pomace extract (GPE) is composed of numerous naturally occurring organic compounds. Accordingly, the inhibitive action of GPE could be attributed to the adsorption of its components on the steel surface. Owing to the complex chemical composition of the GPE, it is rather difficult to assign the inhibitive effect to a particular constituent. The inhibiting mechanism of extract can be viewed at the first stage as a chemisorption of the organic molecules on steel which is followed by formation of polymeric product between organic compounds of extract on the steel surface. The organic depositions observed by SEM on steel surface after long immersion time in presence of the pure inhibiting molecules corroborate this explanation. According to SEM observations, uses of the extract as inhibitor lead to the more significant formation of organic films on steel surface, compared to individual components. In the present study, the inhibition action of two major components, 2-phenylacetaldehyde and hexanal, are studied. Concerning the grape pomace extract, the electrochemical measurements clearly show that its corrosion inhibiting efficiency is better than the one obtained with the pure molecules with approximately the same concentration. Moreover, the analysis of impedance measurements in presence of extract reveals the formation of a film, detected by the appearance of diffusion impedance during time of pretreatment in the vapor phase of extract. This result is supported by SEM analysis of steel surface after a long-time of immersion in the vapor phase of extract.

The inhibitory properties of 2-phenylacetaldehyde are very similar to the extract, which implies that the aldehydes would be the major contributor to the activity. Finally, the anticorrosion properties of the pure molecules are directly related to their chemical structure. Indeed, the inhibition activity of 2-phenylacetaldehyde is better than that of hexanal. The effect of film forming period on the weight loss rate indicated that the extract does not only keep its inhibitive activity for carbon steel under condition of condensation of moisture, but also improves its effectiveness over the long-term test due to synergistic influence of the main compounds which offer an additional protection. According to the quantum chemical calculations, the adsorption of aldehydes is mainly through the –CHO functional group for aromatic aldehydes and through the C=C functional group for unsaturated aldehydes. Noticeably, a number of aldehydes have similar chemical molecular structure, which implies that a series of aldehydes could be seemed as potential contributors for the inhibition performance.

6. Conclusions

Grape pomace extract and two major compounds of 2-phenylacetaldehyde and hexanal act as moderate inhibitors of corrosion of mild steel. Inhibition efficiency values follow the order: grape pomace extract > 2-Phenylacetaldehyde > Hexanal. Individual grape pomace extracts, 2-phenylacetaldehyde and hexanal, act as mixed-type inhibitors of corrosion. In this case, the rate of corrosion is controlled by the cathodic process of reduction of the molecular oxygen from air. It is shown that the inhibition of the atmospheric corrosion of steel after treatment in the vapor of volatile compounds of the grape pomace extract and two major compounds occurs due to the blocking of the metal surface by chemically adsorbed molecules. SEM and AFM results clearly show that GPE and two of the major compounds, 2-phenylacetaldehyde and hexanal, inhibit corrosion of steel by being adsorbed on the metal surface. The better inhibition efficiency of 2-phenylacetaldehyde than hexanal could be explained by the quantum chemical parameters.

References

[1] Abdullah Dar, M. 2011. A review: plant extracts and oils as corrosion inhibitors in aggressive media. Industrial Lubrication and Tribology 63: 227–233.

[2] Zhang, D. -q., Z. -x. An, Q. -y. Pan, L. -x. Gao and G. -d. Zhou. 2006. Volatile corrosion inhibitor film formation on carbon steel surface and its inhibition effect on the atmospheric corrosion of carbon steel. Applied Surface Science 253: 1343–1348.

[3] Zhang, D. -Q., L. -X. Gao and G. -D. Zhou. 2010. Self-assembled urea-amine compound as vapor phase corrosion inhibitor for mild steel. Surface & Coatings Technology 204: 1646–1650.

[4] Sudheer, M. A. Quraishi, E. E. Ebenso and M. Natesan. 2012. Inhibition of atmospheric corrosion of mild steel by new green inhibitors under vapour phase condition. Int. J. Electrochem. Sci. 7: 7463–7475.

[5] Bastidas, D., E. Cano and E. Mora. 2005. Volatile corrosion inhibitors: a review. Anti-Corrosion Methods and Materials 52: 71–77.

[6] Focke, W. W., N. S. Nhlapo and E. Vuorinen. 2013. Thermal analysis and FTIR studies of volatile corrosion inhibitor model systems. Corros. Sci. 77: 88–96.

[7] Quraishi, M. A. and D. Jamal. 2002. Development and testing of all organic volatile corrosion inhibitors. Corrosion 58: 387–391.

[8] Rammelt, U., S. Koehler and G. Reinhard. 2009. Use of vapour phase corrosion inhibitors in packages for protecting mild steel against corrosion. Corros. Sci. 51: 921–925.

[9] Vuorinen, E. and W. Skinner. 2002. Amine carboxylates as vapour phase corrosion inhibitors. British Corrosion Journal 37: 159–160.

[10] Andreev, N. N. and Yu. I. Kuznetsov. 2005. Physicochemical aspects of the action of volatile metal corrosion inhibitors. Russ. Chem. Rev. 74: 685–690.

[11] Andreev, N. N. and Yu. I. Kuznetsov. 2013. Volatile inhibitors of atmospheric corrosion III. Principles and methods of efficiency estimation. Int. J. Corros. Scale Inhib. 2: 39–52.

[12] Ansari, F. A., C. Verma, Y. S. Siddiqui, E. E. Ebenso and M. A. Quraishi. 2018. Volatile corrosion inhibitors for ferrous and non-ferrous metals and alloys: A review. Int. J. Corros. Scale Inhib. 7: 126–150.

[13] Montemor, M. F. 2016. Fostering green inhibitors for corrosion prevention. pp. 107–137. *In*: Hughes A., J. Mol, M. Zheludkevich and R. Buchheit (eds.). Active Protective Coatings. Springer Series in Materials Science, vol. 233. Springer, Dordrecht.

[14] Rathi, P., S. Trikha and S. Kumar. 2017. Plant extracts as green corrosion inhibitors in various corrosive media—a review. World Journal of Pharmacy and Pharmaceutical Sciences 6: 482–514.

[15] Poongothai, N., P. Rajendran, M. Natesan and N. Palaniswamy. 2005. Wood bark oils as vapour phase corrosion inhibitors for metals in NaCl and SO_2 environments. Indian Journal of Chemical Technology 12: 641–647.

[16] Premkumar, P., K. Kannan and M. Natesan. 2008. Thyme extract of Thymus vulgar L. as volatile corrosion inhibitor for mild steel in NaCl environment. Asian Journal of Chemistry 20: 445–451.

[17] Premkumar, P., K. Kannan and M. Natesan. 2008. Natural thyme volatile corrosion inhibitor for mild steel in HCl environment. Journal of Metallurgy and Material Science 50: 227–234.

[18] Chygyrynets', E. and V. Vorobyova. 2014. A study of rape-cake extract as eco-friendly vapor phase corrosion inhibitor. Chemistry & Chemical Technology 8: 235–242.

[19] Vorob'iova, V., O. Chyhyrynets and O. Vasyl'kevych. 2015. Mechanism of formation of the protective films on steel by volatile compounds of rapeseed cake. Mater. Sci. 50: 726–735.

[20] Chyhyrynets', O. E. and V. I. Vorob'iova. 2013. Anticorrosion properties of the extract of rapeseed oil cake as a volatile inhibitor of the atmospheric corrosion of steel. Mater. Sci. 49: 318–325.

[21] Chyhyrynets, O. E., Y. F. Fateev, V. I. Vorobiova et al. 2016. Study of the mechanism of action of the isopropanol extract of rapeseed oil cake on the atmospheric corrosion of copper. Mater Sci. 51: 644–651.

[22] Vorobyova, V., O. Chygyrynets', M. Skiba, I. Kurmakova and O. Bondar. 2017. Self-assembled monoterpenoid phenol as vapor phase atmospheric corrosion inhibitor of carbon steel. International Journal of Corrosion and Scale Inhibition 6: 485–503.

[23] Rockenbach, I., E. Rodrigues, L. Gonzaga, V. Caliari, M. Genovese, A. Gonçalves and R. Fett. 2011. Phenolic compounds content and antioxidant activity in pomace from selected red grapes (*Vitis vinifera* L. and *Vitis labrusca* L.) widely produced in Brazil. Food Chemistry 127: 174–179.

[24] Ostovari, A., S. Hoseinieh, M. Peikari, S. Shadizadeh and S. Hashemi. 2009. Corrosion inhibition of mild steel in 1 M HCl solution by henna extract: A comparative study of the inhibition by henna and its constituents (Lawsone, Gallic acid, α-d-Glucose and Tannic acid). Corros. Sci. 51: 1935–1949.

[25] Gece, G. 2008. The use of quantum chemical methods in corrosion inhibitor studies. Corros. Sci. 50: 2981–2992.

[26] Garai, S., S. Garai, P. Jaisankar, J. Singh and A. Elango. 2012. A comprehensive study on crude methanolic extract of Artemisia pallens (Asteraceae) and its active component as effective corrosion inhibitors of mild steel in acid solution. Corrosi. Sci. 60: 193–204.

[27] Torres, V., R. Amado, C. de Sá, T. Fernandez, C. Riehl, A. Torres and E. D'Elia. 2011. Inhibitory action of aqueous coffee ground extracts on the corrosion of carbon steel in HCl solution. Corrosion Science 53: 2385–2392.

[28] Guo, L., S. Kaya, I. Obot, X. Zheng and Y. Qiang. 2017. Toward understanding the anticorrosive mechanism of some thiourea derivatives for carbon steel corrosion: A combined DFT and molecular dynamics investigation. Journal of Colloid and Interface Science 506: 478–485.

[29] Kaya, S., B. Tüzün, C. Kaya and I. Obot. 2016. Determination of corrosion inhibition effects of amino acids: Quantum chemical and molecular dynamic simulation study. Journal of the Taiwan Institute of Chemical Engineers 58: 528–535.

[30] Pearson, R. G. 1990. Absolute electronegativity, hardness, and bond energies. *In*: Marks, T. J. (ed.). Bonding Energies in Organometallic Compounds, American Chemical Society, Washington.

[31] Chattaraj, P. 2001. Chemical reactivity and selectivity: local HSAB principle versus frontier orbital theory. The Journal of Physical Chemistry A 105: 511–513.

CHAPTER 4

Plant-Based Green Corrosion Inhibition

Need and Applications

Anjali Peter[1] and Sanjay K. Sharma[2],*

1. Introduction

Plant materials and crude species have many active components which are highly attainable and endless wellsprings of a complete field of green inhibitors [1]. They comprise the fruitful reservoirs of constituents which have strong inhibition efficiency and are therefore termed 'Green Inhibitors' [2].

1.1 Properties of Green Inhibitors

Green corrosion inhibitors are biodegradable and do not accommodate heavy elements or other noxious complexes. The helpful application of naturally occurring elements to obstruct the decomposition of metals in the acidic and alkaline environments has been reported [3, 4]. The research attempts to detect real biological materials or biodegradable organic components to be used as effective corrosion inhibitors of a vast estimate of elements have been one of the principal emerging regions in the last two decades [5].

1.2 Economical Worth of Green Inhibitors

Review of natural active components that work as corrosion inhibitors is the subject of worldwide study due to the minor expense and eco-warmth of plant products and is a fast replacement of the unnatural and costly, dangerous organic inhibitors. Plant extracts illustrate an abundance of essential syntheses which have corrosion

[1] Green Chemistry and Sustainability Research Group, Department of Chemistry, JECRC University, Jaipur-303905, India.
[2] Centre of Research Excellence in Corrosion, King Fahd University of Petroleum and Minerals, Dhahran 31261, Kingdom of Saudi Arabia.
* Corresponding author: sk.sharmaa@outlook.com

inhibiting abilities. The yield of these syntheses, as well as the corrosion inhibition techniques, varies largely depending on the character of the plant [6, 3] and its region [7].

1.3 Use of Plant Extracts as Corrosion Inhibitors

The damage of elements in many enterprises, foundations, apparatus, and municipal duties such as public utilities, drink water and sewerage supplies is a complication and causes huge losses. Organic, chemical or a combination of both can inhibit corrosion by either chemisorption on the metal surface or countering with metal ions and implementing a barrier-type precipitate on its facade. Because of the noxious nature or excessive value of some chemicals in operation nowadays as inhibitors, it is imperative to develop environmentally sustainable and reasonable ones.

Natural products can be considered as an absolute support for this purpose. The aqueous extracts from particular divisions of some plants such as Henna, *Lawsonia inermis* [8], *Caricapapaya* [1], *datepalm/phoenix dactylifera* and *Nypa Fruticans Wurmb* have been observed to be good corrosion inhibitors for many metals and alloys.

Plant products are cheaper than organic inhibitors and inexhaustible sources of their components can be easily found. There has been a widespread research on the effectiveness of plant extracts, and they are applied as corrosion inhibitors [9]. Plant extracts including their leaves, barks, seeds, fruits and roots comprise compounds of pure complexes and their corrosion inhibitive properties have been broadly researched. There exist numerous naturally occurring substances that have been reported as anticorrosive inhibitors including Henna, *Nypa fructicans*, natural honey, Opuntia, jojoba oil, *Andrographis paniculata*, Datura, *D. stramonium* and isolated alkaloids like Berberine [2], *Solanum surrattence* [10], Ajowan plant, Seed of Areca catechu [6], *Agremone mexicana*, Black pepper and its piperine [11], Beet root extract, *Bifurcaria bifurcate* [12], Black cumin oil [13], Black pepper extract [14], *Camellia sinensis* [15], *Capparis desidua*, *Citrullus colocynthis* [16], Clove oil [17], *Colocasia esculenta* [18], Fennel [19], Fenugreek leaves and seeds extract [20], Fig leaves extract [21], Karanj seeds extract [22], *Lasienthera africana* [23], *Muntingia calabura*, *Musa sapientum*, Natural honey with Black reddish juice, *Nerium oleander* and *Tecoma stans*, *Ocimum gratissimun* [24], *Ocimum sanctum*, *Palicourea guianensis*, *Piper guinensis* [25], Potato peel extract, *Prunus cerasus*, *Punica granatum*, *Quinine awad*, *Rauvolfia serpentine*, Raphia hookeri, Albomycin, *Sida acuta* [26], *Siparuna guianensis* and *Spirulina platensis* [27]. All these inhibitors are extracted from aromatic species, herbs and medicinal plants.

2. Categories of Major Plants as Green Corrosion Inhibitors

2.1 Guar Gum as a Corrosion Inhibitor

Gums are obtained from plants and they are solids consisting of mixtures of polysaccharides (carbohydrates) which are either water-soluble or absorb water and swell up to form a gel or jelly when placed in water. They are insoluble in oils or organic solvents such as hydrocarbons, ether and alcohol. The mixtures are often complex and on hydrolysis yield simple sugars such as arabinose, galactose, mannose and glucuronic acid. Gums have been found to be good corrosion inhibitors because they form complexes with metal ions and on the metal surface through their functional groups [28].

Gum metal complexes occupy a large surface area, thereby blanketing the surface and shielding the metal from corrosive agents present in the solution. Arbinogalactan, sucrose, oligosaccharides, polysaccharides and glycoprotein are effective in protecting the metal surfaces as they contain oxygen and nitrogen atoms which are the centers of adsorption. Most gums are less toxic, green and eco-friendly. We compiled the list of important Gums which are reported as effective corrosion inhibitors in a comprehensive review article [29].

Table 1. List of natural gums which have been used as corrosion inhibitors.

Metal	Inhibitor	Medium
Mild Steel	Gum Exudate from *Acacia seyal* var. *Seyal*	Drinking Water
Mild Steel	*Anogeissus leiocarpus* gum	0.1 M HCl
Carbon Steel	Guar Gum	--
Aluminum	Gum Arabic	0.1 M NaOH in the absence and present of KI
Aluminum	Exudate Gum from *Pachylobus edulis*	Presence of halide ions in HCl
Al and Mild Steel	Gum Arabic	0.1–0.5 g/L H_2SO_4
Mild steel	Albizia zygia gum	1.1–0.5 g/L H_2SO_4 and HCl
Carbon Steel marked Steel 39, 44, and iron B 500	Locust bean gum (known as carob gum, carob bean)	Sulfuric acid in the presence of chloride ions
Mild steel	*Daniella Olliverri* gum	HCl solution
Al	Exudate gum from *Dacroydes Edulis*	HCl solution
Mild steel	Exudate gum from *Raphiea Hookeri*	H_2SO_4 in the temperature range 30–60°C
Mild Steel	Acacia Gum	HCl and H_2SO_4 solution 0.5 M, 1.0 M & 2.0 M

2.2 *Azadirachta indica* as a Corrosion Inhibitor

Azadirachta indica (commonly recognized as 'Neem') is noteworthy both for its chemical and biological actions. It is one of the most fruitful sources of secondary metabolites in nature. To date, more than 300 natural products have been isolated from different sections of the tree, with new compounds added to the list every year [30].

2.2.1 Chemical Ingredients in Azadirachta indica

Neem is a representative member of the mahogany family, *Meliaceae*. *Azadirachta indica* synthetic constituents hold a variety of 3 or 4 parallel compounds, and it backs these up with 20 or more others that are minor/secondary but nonetheless active in one way or another. In earlier ages, these complexes were associated with a broad assembly of natural products labelled as "triterpenes", or more correctly, "limonoids" [31].

I. Limonoids

Notably, at least nine neem limonoids have shown a capacity to interfere with insects, working on an area of species that is composed of some of the most harmful plagues of agriculture and human health. New limonoids are yet being uncovered in neem, but azadirachtin, salannin, meliantriol and nimbin are the most honorable experienced and, for now at least, suggest being the most meaningful.

II. Azadirachtin

One of the prime active ingredients separated from neem, azadirachtin, has been affirmed to be the tree's prime agent for fighting insects. It does not destroy insects—at least not immediately. Instead, it both revolts and agitates their production and multiplication. Research over the past 20 years has clarified that it is one of the most influential growth regulators and feeding deterrents ever examined. It will shift back or weaken the provision of many varieties of pest insects and some nematodes. In reality, it is obvious that a simple spot of its existence prevents some insects from constantly tender plants.

III. Meliantriol

Another feeding inhibitor, meliantriol, is competent, at exceptionally small frequencies, to breed insects to discontinue feeding. The explanation of its competence to prevent locusts chewing on crops was the primary scientific evidence for neem's acceptable service for insect control on India's crops.

IV. Salannin

A third triterpenoid isolated from neem is salannin. Surveys reveal that this complex also effectively inhibits feeding but does not influence insect moults. The migratory locust, California red scale, striped cucumber beetle, houseflies and the Japanese beetle have been strongly discouraged in both laboratory and field assessments.

V. Nimbin and Nimbidin

Two more neem components, nimbin, and nimbidin, have been discovered to have antiviral action. They affect potato virus X, vaccinia virus and fowl pox virus. They could conceivably introduce a procedure to deal with these and new viral illnesses of crops and stock [29].

2.2.2 Overview of the Application of Azadirachta indica as an Inhibitor

The uses of Neem as a corrosion inhibitor for different metals is described in our published review paper [29]. Table 2 presents examples of the use of *Azadirachta indica (AZI)* as a corrosion inhibitor.

Table 2. Summary of the use of *Azardirachta indica* (AZI) as a corrosion inhibitor.

Metal	Medium	Inhibitor	Method	Findings
Mild steel	H₂SO₄	Leaves, root and seeds extracts of AZI	Weight loss and asometric techniques	Freundlich adsorption isotherm
Mild steel	2 M HCl and 1 M H₂SO₄	AZI	Kinetics activation parameters, temp. 30°C and 60°C	Responsible for the inhibiting action of the extracts
Al	0.5 M HCl	AZI and iodide ions as additive	Potentiodynamic Polarization technique	Freundlich adsorption isotherm
Mild steel	2 M HCl and 1 HCl	AZI extracts	Gas-volumetric Technique	Langmuir isotherm mixed inhibitor
Al	HCl	AZI	Weight loss thermometric and hydrogen evolution techniques, temp 30–40°C	Freundlich, Temkin and Flory-Huggins adsorption isotherms
Al	0.5 M HCl	Fruit of AZI	Weight loss thermometric method	High corrosion inhibition efficiency
Mild steel	HNO₃	AZI mature leaf extract	Gravimetric techniques	Inhibition efficiency varies with concentration
Carbon steel	1.0 M HCl	UAE Neem extract	Weight loss method	Temkin adsorption isotherm
Stainless steel	HCl solution, Tetraoxosulphate (IV) acid solution, Trioxonitrate (V) acid solution	Plant extract	Weight loss method	Good inhibition efficiency
Mild steel	H₂SO₄	Ethanol extracts of seeds and leaves of AZI	Gravimetric, Gasometric and IR methods	Flory Huggins adsorption isotherm
Mild steel	0.5 M HCl and H₂SO₄ acids	Extract of AZI leaf	Weight loss	Appreciable performance

3. Nanocomposites as Green Inhibitors

Nanocomposites are solid materials that have multiple phase domains and at least one of these domains has a nanoscale dimension. The materials can have novel chemical and physical properties that depend on the outer surface and internal characteristics of the component matter. Readily available natural material and polymers such as carbohydrates, fats and proteins make hard composites as calcite bones, shell and wood; these are made by mixing two or more particles and at least one of them is in the nanoscale size. Nanoscale science and technology research is progressing with the use of a combination of atomic scale characterization and detailed modelling.

Toyota Central Research Laboratories published work on a Nylon-6 nanocomposite and these material showed thermal and mechanical properties resistance [32].

Nanotechnology acquires futuristic approach and accordingly the US federal funding for R&D in nanotechnology has increased substantially since the inception of the National Nanotechnology Initiative (NNI) and raising its funding from $464 million in 2001 to $982 million in 2005 [33]. The 2020 federal budget provides more than $1.4 billion for the NNI.

Nanomaterials accommodate fortified proficiency as much as their high countenance ratios. The productivity of a nanocomposite is highly dependent on the size and the degree of blending between the two phases. The nanoparticle and coherence of the particle–matrix interface have a critical impact on deciding the standardized gain of the nanocomposite. Without accurate overthrow or blending, the nanomaterial will not support enhanced standardized parameters over that of sanctioned composites. In fact, an indisposed nanomaterial may demolish the standard properties. For example, good adhesion at the interface will improve properties such as interlaminar shear strength, delimitation resistance, fatigue and corrosion resistance [33]. The types and various applications of nanocomposites are described in the scheme below:

Types of Nanocomposites

Importance of Particles Properties

Particle size and shape can significantly influence the nanostructures properties such as stability, chemical reactivity, texture and porosity. Interactions at phase interfaces of nanoparticles can largely improve the material's properties. In this context, the surface area/volume ratio of reinforcement materials employed in the preparation of nanocomposites is crucial to the understanding of their structure-property relationships.

Ceramic Matrix Nanocomposites (CMNC)

Example is Al_2O_3/SiC system, in which incorporation of high strength nanofibers into ceramic matrices has allowed the preparation of advanced nanocomposites with high toughness and superior failure characteristics compared to the sudden failures of ceramic materials

Metal Matrix Nanocomposites (MMNC)

Materials consisting of a ductile metal or alloy matrix in which some nanosized reinforcement material is implanted. These materials combine metal and ceramic features, i.e. ductility and toughness with high strength and modulus.

Polymer Matrix Nanocomposites (PMNC)

Materials that consist of polymeric substances like carbohydrades, lipids and protein and are prepared by the mixing of two or more particles.

Applications

* Drug delivery systems

* Anti-corrosion barrier coatings

*New fire retardant materials

* Superior strength fibers and films

* Lubricants and scratch-free paint

* UV protection gels

A nanocomposite plays a major role as a green corrosion inhibitor due to its compatibility, small size, good adsorption and binding capacity. Table 3 shows examples of plant-based extracts and their uses in corrosion inhibition and other potential applications.

Table 3. Summary of plant-based nanocomposites and their application in corrosion inhibition.

Nanocomposite	Plant part	Applied technique	Results	Applicability	Reference
Graphene/Silver (G/Ag) and Graphene/Gold (G/Au)	*Xanthium strumarium* leaf plant extract	UV-Vis, TEM, SEM, XPS, XRD, Raman spectroscopy and EDX	G/Ag and G/Au nano-composites became eco-friendly using *Xanthium strumarium* leaf extract.	The produced graphene/ metal nano composites may be less-toxic, biocompatible and useful for bio-applications.	[34]
Hydrophilic Composites. Poly(Ethylene Glycol) Methyl Ether Acrylate-PPEGMEA	a plant virus protein nanotube, tobacco mosaic green mild virus (TMGMV)	------	MGMV was an effective reinforcing agent at low concentrations, increasing the storage and compressive modulus when compared to neat PPEGMEA hydrogels.	Polyvinyl alcohol (PVA)/ TMGMV composite films were studied. TMGMV was found to have a similar reinforcing effect on the films. Rod-shaped viruses may represent a broad-based reinforcing agent for aqueous systems in the future.	[35]
Styrene and clay organo-modifier	Functionalized triglyceride, such as acrylated epoxidized soybean oil, maleinized acrylated epoxidized soybean oil, and soybean oil pentaerythritol maleates	X-ray data and TEM	The flexural modulus increased 30% at only 4 vol % clay content.	----	[36]

Table 3 Contd. ...

...*Table 3 Contd.*

Nanocomposite	Plant part	Applied technique	Results	Applicability	Reference
Green synthesis of silver NPs (Ag NPs)	leaf extracts of *Azadirachta indica*	microwave assisted synthesis of AgNPs	Reported as a novel nano-composite to adsorb hazardous dyes.	Ag-nanocomposite, as an adsorbent, helped in achieving about 97% removal of crystal violet dye.	[37]
Organophilic Mont-morillonite Produced Triglyceride–Clay Nano-composites	Epoxidized plant oils	----	A nano-composite with homogeneous structure was obtained, in which silicate layers of the clay were intercalated and randomly distributed in the polymer matrix.	----	[38]
A Reduced Graphene Oxide (RGO)/ Fe$_3$O$_4$ based Nano-composite with Palladium Nano-particles (Pd NPs)	*Withania* coagulants leaf extract.	XRD, FE-SEM, EDS, VSM, TEM, FT-IR and UV–vis.	Reducing and stabilizing agent.	Its catalytic activity has been tested for the reduction of 4-nitrophenol (4-NP) in water at room temperature.	[39]
Ag/TiO$_2$ nano-composite	extract of leaves of *Euphorbia heterophylla*	FE-SEM, EDX, FT-IR, XRD and UV–vis	It is found to be an effective catalyst for reduction of various dyes.	Used as a catalyst for many dyes.	[40]
Synthesis of anti-microbial silver nano-particles	callus and leaf extracts from saltmarsh plant, *Sesuvium portulacastrum* L.	XRD, TEM, FTIR	The silver nanoparticles were observed to inhibit clinical strains of bacteria and fungi. The antibacterial activity was more distinct than antifungal activity.	The antimicrobial activity was enhanced when polyvinyl alcohol was added. Tissue culture-derived callus extract from the coastal saltmarsh species was used for the synthesis of antimicrobial silver.	[41]

4. Conclusions

Application of natural products as corrosion inhibitors is the subject of worldwide study due to the minor expense and eco-warmth of plant products; this approach is also a fast replacement of the unnatural and costly, dangerous organic inhibitors. Green inhibitors are the natural warrior against corrosion with limited weapons. Commercialization is still in process to introduce green inhibitors, but without blending them with some strong basic substance, it will not be possible to keep these inhibitors intact. A nanocomposite is the final answer for long stability and strong capacity for holding. Nanocomposites are the combination of two nanomaterials (namely, metals + green inhibitor). A nanocomposite leads to good results with future applicability for the prevention of corrosion.

References

[1] Okafor, P. C. and E. E. Ebenso. 2007. Inhibitive action of Carica papaya extracts on the corrosion of mild steel in acidic media and their adsorption characteristics. Pigment & Resin Technology 36: 134–140.

[2] Sangeetha, M., S. Rajendran, T. S. Muthumegala and A. Krishnaveni. 2011. Green corrosion inhibitors—An overview. Zastita Materijala 52: 3–19.

[3] Okafor, P. C., M. E. Ikpi, I. E. Uwah, E. E. Ebenso, U. J. Ekpe and S. A. Umoren. 2008. Inhibitory action of Phyllanthus amarus extracts on the corrosion of mild steel in acidic media. Corrosion Science 50: 2310–2317.

[4] Okafor, P. C., V. I. Osabor and E. E. Ebenso. 2007. Eco-friendly corrosion inhibitors: inhibitive action of ethanol extracts of Garcinia kola for the corrosion of mild steel in H_2SO_4 solutions. Pigment & Resin Technology 36: 299–305.

[5] Sharma, S., A. Chaudhary and R. V. Singh. 2008. Gray chemistry verses green chemistry: Challenges and opportunities. Rasayan Journal of Chemistry 1: 68–92.

[6] Vinod Kumar, K. P., M. Sankara Narayanan Pillai and G. Rexin Thusnavis. 2011. Green corrosion inhibitor from seed extract of Areca catechu for mild steel in hydrochloric acid medium. J. Mater. Sci. 46: 5208–5215.

[7] Ogan, A. U. 1971. The basic constituents of the leaves of Carica papaya. Phytochemistry 10: 2544–2547.

[8] El-Etre, A. Y., M. Abdallah and Z. E. El-Tantawy. 2005. Corrosion inhibition of some metals using lawsonia extract. Corrosion Science 47: 385–395.

[9] Raja, P. B. and M. G. Sethuraman. 2008. Natural products as corrosion inhibitor for metals in corrosive media—A review. Materials Letters 62: 113–116.

[10] Sharma, M. K., A. K. Sharma and S. P. Mathur. 2012. Solanum surrattence as potential corrosion inhibitor. ISRN Corrosion 2012: 1–5.

[11] Dahmani, M., R. Touzani, S. Al-Salem, B. Hammouti and A. Bouyanzer. 2010. Corrosion inhibition of C38 steel in 1 M HCl: A comparative study of black pepper extract and its isolated piperine, International Journal of Electrochemical Science 5: 1060–1069.

[12] Abboud, Y., A. Abourriche, T. Ainane, M. Charrouf, A. Bennamara, O. Tanane and B. Hammouti. 2009. Corrosion inhibition of carbon steel in acidic media by Bifurcaria bifurcata extract. Chemical Engineering Communications 196: 788–800.

[13] Abdallah, M., S. O. Al Karanee and A. A. Abdel Fatah. 2010. Inhibition of acidic and pitting corrosion of nickel using natural black cumin oil. Chemical Engineering Communications 197: 1446–1454.

[14] Quraishi, M. A., D. K. Yadav and I. Ahamad. 2009. Green approach to corrosion inhibition by black pepper extract in hydrochloric acid solution. TOCORRJ 2: 56–60.

[15] Loto, C. 2011. Inhibition effect of Tea (Camellia Sinensis) extract on the corrosion of mild steel in dilute sulphuric acid. Journal of Materials and Environmental Science 2.

[16] Chauhan, R., U. Garg and R. K. Tak. 2011. Corrosion inhibition of aluminium in acid media by Citrullus colocynthis extract. E-Journal of Chemistry 8: 85–90.

[17] Abdallah, M., S. O. Al Karanee and A. A. Abdel Fattah. 2009. Inhibition of the corrosion of nickel and its alloys by natural clove oil. Chemical Engineering Communications 196: 1406–1416.

[18] Eddy, N. O. and P. A. P. Mamza. 2009. Inhibitive and adsorption properties of ethanol extract of seeds and leaves of Azadirachta indica on the corrosion of mild steel in H_2SO_4, port. Electrochim. Acta 27: 443–456.

[19] Lahhit, N., A. Bouyanzer, J. -M. Desjobert, B. Hammouti, R. Salghi, J. Costa, C. Jama, F. Bentiss and L. Majidi. 2011. Fennel (Foeniculum vulgare) essential oil as green corrosion inhibitor of carbon steel in hydrochloric acid solution, port. Electrochim. Acta 29: 127–138.

[20] Noor, E. 2008. Comparative study on the corrosion inhibition of mild steel by aqueous extract of fenugreek seeds and leaves in acidic solutions. J. Eng. Appl. Sci. 3.

[21] Ibrahim, T. and M. Abouzour. 2011. Corrosion inhibition of mild steel using fig leaves extract in hydrochloric acid solution. International Journal of Electrochemical Science 6: 6442–6455.

[22] Singh, A., I. Ahamad, V. Singh, and M. A. Quarishi. 2011. Inhibition effect of environmentally benign Karanj (Pongamia pinnata) seed extract on corrosion of mild steel in hydrochloric acid solution. J. Solid State Electrochem. 15: 1087–1097.

[23] Njoku, V. O., E. E. Oguzie, C. Obi and A. A. Ayuk. 2014. Baphia nitida leaves extract as a green corrosion inhibitor for the corrosion of mild steel in acidic media. Advances in Chemistry 2014: 1–10.

[24] Eddy, N. O., S. A. Odoemelam and I. N. Ama. 2010. Ethanol extract of Ocimum gratissimum as a green corrosion inhibitor for the corrosion of mild steel in H_2SO_4. Green Chemistry Letters and Reviews 3: 165–172.

[25] Ebenso, E., N. Eddy and A. O. Odiongenyi. 2008. Corrosion inhibitive properties and adsorption behaviour of ethanol extract of Piper guinensis as a green corrosion inhibitor for mild steel in H_2SO_4. Afr. J. Pure Appl. Chem. 2.

[26] Umoren, S. A., U. M. Eduok, M. M. Solomon and A. P. Udoh. 2016. Corrosion inhibition by leaves and stem extracts of Sida acuta for mild steel in 1 M H_2SO_4 solutions investigated by chemical and spectroscopic techniques. Arabian Journal of Chemistry 9: S209–S224.

[27] Kamal, C. and M. G. Sethuraman. 2012. Spirulina platensis—A novel green inhibitor for acid corrosion of mild steel. Arabian Journal of Chemistry 5: 155–161.

[28] Eddy, N. O., P. Ameh, E. Gimba Casmir and E. E. Ebenso. 2011. GCMS studies on Anogessus leocarpus (Al) gum and their corrosion inhibition potential for mild steel in 0.1 M HCl. Int. J. Electrochem. Sci. 5815–5829.

[29] Peter, A., I. B. Obot and S. K. Sharma. 2015. Use of natural gums as green corrosion inhibitors: an overview. Int. J. Ind. Chem. 153–164.

[30] Palanisamy, K. and P. Kumar. 1997. Effect of position, size of cuttings and environmental factors on adventitious rooting in neem (*Azadirachta indica* A. Juss). Forest Ecology and Management 98: 277–280.

[31] Bhola, S. M., F. M. Alabbas, R. Bhola, J. R. Spear, B. Mishra, D. L. Olson and A. E. Kakpovbia. 2014. Neem extract as an inhibitor for biocorrosion influenced by sulfate reducing bacteria: A preliminary investigation. Engineering Failure Analysis 36: 92–103.

[32] Usuki, A., M. Kawasumi, Y. Kojima, A. Okada, T. Kurauchi and O. Kamigaito. 1993. Swelling behavior of montmorillonite cation exchanged for ω-amino acids by ε-caprolactam. J. Mater. Res. 8: 1174–1178.

[33] National Nanotechnology Initiative. 2020. Federal Budget. https://www.nano.gov.

[34] Wen, J., B. K. Salunke and B. S. Kim. 2017. Biosynthesis of graphene-metal nanocomposites using plant extract and their biological activities. J. Chem. Technol. Biotechnol. 92: 1428–1435.

[35] Zheng, Y., M. L. Dougherty, D. Konkolewicz, N. F. Steinmetz and J. K. Pokorski. 2018. Green nanofillers: Plant virus reinforcement in hydrophilic polymer nanocomposites. Polymer 142: 72–79.

[36] Lu, J., C. K. Hong and R. P. Wool. 2004. Bio-based nanocomposites from functionalized plant oils and layered silicate. J. Polym. Sci. B Polym. Phys. 42: 1441–1450.

[37] Satapathy, M. K., P. Banerjee and P. Das. 2015. Plant-mediated synthesis of silver-nanocomposite as novel effective azo dye adsorbent. Appl. Nanosci. 5: 1–9.

[38] Uyama, H., M. Kuwabara, T. Tsujimoto, M. Nakano, A. Usuki and S. Kobayashi. 2003. Green nanocomposites from renewable resources: plant oil–clay hybrid materials. Chem. Mater. 15: 2492–2494.

[39] Atarod, M., M. Nasrollahzadeh and S. M. Sajadi. 2015. Green synthesis of a Cu/reduced graphene oxide/Fe$_3$O$_4$ nanocomposite using Euphorbia wallichii leaf extract and its application as a recyclable and heterogeneous catalyst for the reduction of 4-nitrophenol and rhodamine B. RSC Adv. 5: 91532–91543.

[40] Atarod, M., M. Nasrollahzadeh and S. Mohammad Sajadi. 2016. Euphorbia heterophylla leaf extract mediated green synthesis of Ag/TiO$_2$ nanocomposite and investigation of its excellent catalytic activity for reduction of variety of dyes in water. Journal of Colloid and Interface Science 462: 272–279.

[41] Nabikhan, A., K. Kandasamy, A. Raj and N. M. Alikunhi. 2010. Synthesis of antimicrobial silver nanoparticles by callus and leaf extracts from saltmarsh plant, Sesuvium portulacastrum L. Colloids and Surfaces. B, Biointerfaces 79: 488–493.

CHAPTER 5

Corrosion Protection Using Organic and Natural Polymeric Inhibitors

Ebru Saraloğlu Güler

1. Introduction: Corrosion Protection Methods

One of the results of electrochemical reactions between metals and their environment is corrosion, ending up with the deterioration of a metal. Corrosion must be considered as an important design parameter because it comes with a price. The annual global direct cost of corrosion was estimated to be $2.5 trillion (2016), which is roughly 3.5% of the world's gross domestic product (GDP) according to NACE International [1].

The corrosion protection basically consists of barrier and cathodic protection. A potential is introduced on the protected metal during cathodic protection either by sacrificing another metal (sacrificial anode) or applying direct current to reverse the reaction. Painting, coating and corrosion inhibitors are also among the barrier protection methods. Coatings can be categorized as organic and inorganic coatings and it is applied before the exposure to corrosive environment (Fig. 2a). The nature of the binder determines whether the coating is organic or inorganic. Coating for corrosion protection can be applied by several methods like dip coating [2], spin coating [3], drying followed by simple immersion [4], micropipette deposition [5] and electroplating [6].

Organic or inorganic corrosion inhibitors have been widely used due to the advantage of *in situ* intervention. The performance of an inhibitor depends on the corrosive environment like dissolved salts, pH, temperature and concentration [7]. Polymers are mentioned as good corrosion inhibitors according to the ability of forming complexes of their functional groups with metal ions and depositing on the metal surface. Organic polymers can be either synthetic or natural and both of them can be used for corrosion protection (Fig. 1). Synthetic organic compounds are eliminated since they are expensive, toxic to humans and environment [8, 9].

Department of Mechanical Engineering, Başkent University, 06790 Etimesgut, Ankara, Turkey.
Email: esguler@baskent.edu.tr

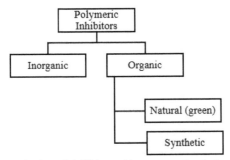

Fig. 1. Types of polymeric inhibitors with respect to chemical composition.

Toxicity and environmental effects of the inhibitors are critical since they are added into the corrosive environment. This chapter focuses on natural (green) polymeric inhibitors. Green inhibitors are attractive because of the properties of chemical and thermal stability, non-flammability, low vapor pressure, and high ionic conductivity [10]. Recognizable examples of natural polymers are proteins, starch, cellulose, hemicelluloses, chitin, casein and silk that are produced by living organisms in nature, mainly animals and plants [11, 12].

2. Adsorption Mechanisms of Inhibitors

Chemical compounds that are called corrosion inhibitors are added to the corrosive environment in order to control or decrease the level of corrosion of metallic surfaces [13]. The corrosive environment can be any kind of system like cooling systems, refinery units, chemicals, oil and gas production units and boilers [14]. Simply a film is generated by the inhibitor and acts as a barrier between the surface and the corrosive environment. The metal is protected from corrosion with the help of decreasing/increasing anodic and/or cathodic reaction, by decreasing the diffusion rate related with the transference of the reactant to the metal surface and by decreasing the electrical resistance of the metal surface [14]. Therefore, inhibitors can be classified as cathodic, anodic and mixed inhibitors [15]. Mixed inhibitors including organic compounds hinder both of the cathodic and anodic reactions that results in no significant change in the corrosion potential [15]. Justicia gendarussa extract (JGPE) was stated as a mixed-type inhibitor to protect mild steel from HCl medium according to the polarization studies [9]. In addition, benzotriazole (BTAH) that is an example of mixed type inhibitor was declared to be the best inhibitor for copper and its alloys [16].

The adsorption capability of the surfactants on the metal surface indicates the corrosion inhibition efficiency [15]. The way of the inhibitors or surfactants to inhibit the corrosion in aqueous solution is physical (electrostatic) or chemical adsorption (chemisorption) of the surfactant molecules onto the corroding metal surface [17, 18]. Schematic representation of physisorption and chemisorption can be seen in Figs. 2b and c, respectively. Multiple layers may be observed in physical adsorption whereas monolayers are created on the surface during chemisorption (Fig. 2b). The adsorption process commonly depends on the electrolyte type, metal's surface

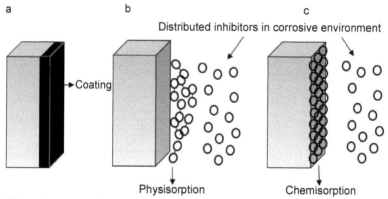

Fig. 2. Schematic representation of (a) coating prior to corrosive environment contact and film formation by (b) physical adsorption and (c) chemical adsorption in the corrosive environment by inhibitor addition.

charge and the inhibitor structure [15]. Charge transfer occurs between phases during chemical adsorption, thus the charge of the adsorbed surfactant affects the efficiency of the inhibitor [19]. The type of the adsorption mechanism (whether chemical or physical) of the inhibitor on the metal surface is expressed by the standard free adsorption energies [15]. High and low free adsorption energies (–100 kJ/mol and –20 to –40 kJ/mol) were calculated for chemical and physical adsorption conditions, respectively [15, 20]. It is well known that the negative free energy is a sign of easy adsorption [21, 22]. Gibbs equation can be applied to Freundlich isotherm so the adsorption free energy can be calculated [21, 22].

3. Protection Performance Measurement Methods

Protection performances of the inhibitors were evaluated by corrosion tests like weight loss test, EIS, salt spray test, cathodic disbanding, potentiodynamic polarization and SEM observations. The simplest way of evaluation is *weight loss test* that consists of the immersion of the two testing metals in the corrosive environments, both including (maybe at several concentrations) and not including the inhibitor for a specific time and weighing the specimens. Weight loss versus immersion time graphs are preferred in these investigations to compare the result of different studies. The inhibition efficiency can be easily calculated by equation 1 [23, 24]:

$$IE\% = \frac{W_{corr} - W_{corr(inh)}}{W_{corr}} \times 100 \tag{1}$$

where *IE* is inhibitor efficiency, $W_{corr(inh)}$ and W_{corr} are the weights of the specimens that are immersed in the including and non-including inhibitor corrosive environment after a specific time, respectively.

Potentiodynamic polarization is another way to evaluate the progress of the corrosion. The polarization scans can be either cathodic or anodic due to the shift of the potential from corrosion potential (E_{corr}) to negative or positive direction, respectively. The corrosion potential and the corrosion current density (I_{corr}) can be determined by the intersection of anodic and cathodic curves [25]. The relationship of current and potential is also called Tafel diagram. The electroplating parameters (I_{corr}, E_{corr}, cathodic Tafel slope—β_c and anodic slope—β_a) can be recorded from Tafel plot and the corrosion current density can be used to determine inhibition efficiency (*IE%*) as stated in equation 2 [23, 24, 26]:

$$IE\% = \frac{I_{corr} - I_{corr(inh)}}{I_{corr}} \times 100 \qquad (2)$$

where *IE* is inhibitor efficiency, $I_{corr(inh)}$ and I_{corr} are the corrosion current densities in the presence and absence of the organic additives, respectively.

The decrease in the current density was stated as a sign of inhibiting effect [27]. On the other hand, the inhibitor efficiency can also be calculated by equation 3 from polarization resistance values (R_p) [26, 28]:

$$IE_{Rp}\% = \frac{R_p - R'_p}{R'_p} \times 100 \qquad (3)$$

where R'_p and R_p are the polarization resistances with and without organic inhibitor in the corrosive environment, respectively.

The polarization resistance is determined by the slope of the potential versus current graph at the corrosion potential by equation 4 [29]. It was also mentioned that R_p increases due to the adsorption of the inhibitor on the metal surface [26].

$$R_p = \left(\frac{dE}{dI} \right)_{E_{corr}} \qquad (4)$$

Furthermore, *electrochemical impedance spectroscopy (EIS)* method is also used for the evaluation of corrosion progress. The method analyzes the electrical current response of the system depending on the frequency [30]. The charge transfer resistance (R_t) corresponds to the diameter of the circle that can be obtained from the EIS data Nyquist plot [23]. The calculation of the inhibition efficiency from charge transfer resistance is shown by the below equation [23, 26]:

$$IE_R\% = \frac{R'_t - R_t}{R'_t} \times 100 \qquad (5)$$

where R'_t and R_t are the values of charge transfer resistance with and without inhibitor.

4. Organic Inhibitors

The coatings applied to the corroding surfaces before the corrosion exposure are not permanent, not cost-effective and they are prone to abrasion and mechanical damage. Therefore, alternative protection methods are more attractive [31]. As it is mentioned, corrosion inhibitor addition is a promising method to control the corrosion. Synthetic organic polymeric inhibitors have been recently used to decrease the deterioration effect of alloys caused by corrosion. Synthetic inhibitors are artificially synthesized to mimic natural inhibitors as an alternative [32]. Organic compounds containing nitrogen, sulphur and oxygen have been commonly recognized to inhibit corrosion [33]. Nitrogen, sulphur and oxygen atoms ensure the adsorption of organic inhibitors which is directly reflected to the efficiency of the inhibitor [34]. Organic inhibitor efficiencies were studied for metals against different mediums. For instance, the inhibition efficiencies of "sodium alkyl sulfate", "sodium cumene sulfonate" and "sulfonic acid" were listed in an increasing order for the aluminum corrosion in hydrochloric acid (HCl) [35]. Triazole based inhibitors such as 3,4-dichloro-acetophenone-O-1'-(1',3',4'-triazolyl)-methaneoxime (4-DTM) and 2,5-dichloro-acetophenone-O-1'-(1',3',4'-triazolyl)-methaneoxime (5-DTM) have been reported as good corrosion inhibitors for steel in acidic media [36, 37].

Temperature of the corrosive environment and the amount of the inhibitor are the main parameters that affect the corrosion inhibition efficiency. The inhibition efficiency was increased by increasing inhibitor concentration and decreasing the temperature in the case of using "N-1-naphthylethylenediamine dihydrochloride monomethanolate" (N-NEDHME) as corrosion inhibitor for carbon steel in 2 M H_3PO_4 at 303–328 K [33]. Similarly, increasing temperature decreased the inhibition efficiency of the organic compound used for corrosion of mild steel in 0.5 M sulfuric acid [38]. The same effect of temperature on inhibition efficiency was also observed in corrosion inhibition of "1, 1', 5 ,5'-tetramethyl-1H, 1'H-3, 3'-bipyrazole" for C38 steel in hydrochloric acid [39]. Moreover, increasing temperature led to an increase in the corrosion current density [15, 37]. Thus, it can be concluded that temperature effect overwhelmed the inhibitor effect. Regarding the temperature, increasing additive (N, N-bis (salicylidene)-2-hydroxy-1,3-propanediamine and N,N-bis(2-hydroxyacetophenylidene)-2-hydroxy-1,3-propanediamine) concentration decreased the corrosion rate and increased inhibition efficiency together with the amount of surface coverage during the corrosion experiment of mild steel in hydrochloric acid [28]. The properties of the corrosive medium, if it is alkaline, neutral acidic or even the type of the acid, affects the inhibition efficiency. For instance, the inhibition efficiency of 4-MTHT was better in 1 M hydrochloric acid than in 0.5 M sulfuric acid and the corresponding efficiency values were 99% and 80%, respectively [37]. Inhibition efficiency values were 20%, 45.1% and 83.5% when chitosan is added to the solutions of "acid chloride", "acid chloride and sulphate mixture" and "acid sulphate", respectively [40].

5. Natural Inhibitors

The use of some organic inhibitors which have the disadvantage of toxicity is limited and need for new environmentally friendly inhibitors has come up [36]. Although the organic compounds, especially those with N, S and O other than natural polymers, have excellent efficiency of corrosion inhibition, they are rejected due to their high cost and/or toxicity to environment [14]. Greenness is one of the major requirements during the selection of the corrosion inhibitor [18]. Additional substances such as organic solvents, environmental hazards, unsafe situations, and high amount of waste are in conflict with sustainability. Therefore, "green corrosion inhibitors", in other words "natural inhibitors", shown under organic polymers in Fig. 1 have been attracting attention due to the toxic effect of some organic and inorganic corrosion inhibitors to the environment [41]. In addition, it is a fact that natural substances have attracted great interest nowadays because they are economically favorable and also environmentally friendly, so they are also called "eco-friendly inhibitors". Similarly, natural polymers among polymeric inhibitors have also been widely reported for corrosion inhibition of metallic materials. Natural inhibitors have the advantages of non-toxicity, renewability, availability and biodegradability. There are many naturally existing assets revealed to have anticorrosion effect and they are extracted from their own aromatic species, herbs and medical plants [14]. Plants like seeds, fruits, leaves and flowers have been broadly studied as natural inhibitors. Natural inhibitors consist of an active structure part and this part varies from one plant to another [42].

Inhibition efficiencies (*IE*) of natural inhibitors are compared in Table 1. Addition of 10^{-4} M chitosan in acidic solution led to better corrosion protection compared to the same amount of N-NEDHME and 5-DTM as can be seen in Table 1. In addition, Jojoba oil generated approximately 100% corrosion protection besides the advantages of being natural (Table 1).

Table 1. Inhibition efficiencies of organic and natural inhibitors studied for steel corrosion in acidic medium.

Inhibitor type	Corrosive medium	Ref. #	Concentration	IE (%)
Organic inhibitor				
N-NEDHME	H$_3$PO$_4$	[33]	10^{-2} M	87.9
			10^{-3} M	76.5
			10^{-4} M	64.8
			10^{-5} M	51.2
4-DTM			1×10^{-5} M	27.5
			3.2×10^{-5} M	73.0
			1×10^{-4} M	85.8
			3.2×10^{-4} M	98.4
	HCl	[36]	1×10^{-3} M	98.7
5-DTM			1×10^{-5} M	23.5
			3.2×10^{-5} M	51.8
			1×10^{-4} M	85.3
			3.2×10^{-4} M	95.1
			1×10^{-3} M	98.4
Natural Inhibitor				
Chitosan	HCl	[43]	10^{-8} M	40
			10^{-7} M	50
			10^{-6} M	64
			10^{-5} M	88
			10^{-4} M	86

Table 1 Contd. ...

...Table 1 Contd.

Inhibitor type	Corrosive medium	Ref. #	Concentration	IE (%)
Natural Inhibitor				
Salivia aucheri mesatlantica oil	H_2SO_4	[23]	0.25 g/l	78.3
			0.5 g/l	79.2
			1 g/l	82.1
			1.5 g/l	84.9
			2 g/l	86.1
Jojoba oil	HCl	[44]	0.515 g/l	98
			0.386 g/l	94
			0.129 g/l	88
			0.005 g/l	58
Caffeine	HCL (35°C)	[45]	10^{-6} M	79
			10^{-5} M	87
			10^{-4} M	88
			10^{-3} M	90

Natural polymers can be categorized in three groups according to their chemical structure: polysaccharides that are biopolymers [18], proteins and polyesters [46]. The sources of polysaccharides are plants, animals and microorganisms so they are extensively available in nature [46]. For instance, low carbon steel was protected by bacterial exopolysaccharide coatings produced by the microorganisms that led to decrease in ionic diffusion rate [47]. The polysaccharides derived from plants like cellulose [48], starch [49–51], and gum [52, 53] are widely used as corrosion inhibitors. The fruits of *argemone mexicana, the leaves, flowers and fruits of calotropis procera, garlic, carrot, castor seed, lignin* have been demonstrated as good corrosion inhibitors with an efficiency of more than 60% in 3% NaCl for mild steel [54]. Moreover, *tobacco, castor seed, black pepper, lignin, soya bean and margosa* protected mild steel at an amount of 70% from sulfuric acid and range of 65%–87.5% from hydrochloric acid [54]. Other natural corrosion inhibitors are *mimosa tannin* [55], *carboxymethyl cellulose* (CMC) [56, 57] and *chitosan* [43] that are used extensively to protect iron-carbon alloys in acid solutions. Chitosan is known as an important biopolymer. *Natural gums* are also used as corrosion inhibitors since they have unique chemical composition [18]. Ionic and uncharged polymers are the main sources such as agar-extracted from *seaweeds and guar gum* from guar bean seeds, respectively [18]. *Gum arabic* also protects metals from corrosive environments including both alkaline and acidic mediums [18]. Gum arabic is a complex polysaccharide collected as a mixture of calcium, magnesium and potassium salt [58]. Other gums used as corrosion inhibitors for iron-carbon alloys are *exudate gums* [52, 59] and *anogessus Leocarpus (Al) gum* [60].

6. Effects of Parameters of Inhibitor and Corrosive Medium on the Inhibition Efficiency

The *type* and the *concentration* of the corrosion inhibitor, *acidity* and the *temperature* of the corrosive medium are the most important parameters that affect the corrosion inhibition efficiency. The corrosion efficiencies were calculated by the main methods as described in "Part 3: Protection Performance Measurement Methods". Increasing *mimosa tannin* concentration in sulfuric acid at different pH values (1, 2, 3) and decreasing pH from 1 to 3 during keeping the inhibitor concentration constant, increased the inhibition efficiency for low carbon steel [55]. In addition, the highest inhibitor efficiency, 87.3%, was reached when inhibitor, mimosa tannin, concentration was 5×10^{-2} mol/l and pH was 1 [55]. Increasing inhibitor (*guar gum*) concentration from 250 to 1500 ppm in 1 M sulfuric acid led to an increase in the inhibition efficiency from 75.6% to 93.9% due to the weight loss measurement by cause of surface coverage increment on carbon steel [53]. On the other hand, increasing inhibitor, *natural honey*, concentration for steel corrosion protection in saline water decreased the performance of the inhibitor due to fungi growth in the medium [61]. The corrosion inhibition effects of *hydroxypropyl cellulose* (HPC), *glucose* and *gellan gum* for cast iron were studied in 1 M HCL at different concentrations and HPC was determined as the one with highest inhibition efficiency [20].

Temperature of the corrosive environment has critical effects on the corrosion inhibition efficiency. Therefore, thermal stability of the inhibitors is critical especially for the parts that are exposed to high temperatures. Biopolymer inhibitors were compared according to their thermal stabilities and they can be listed as *CMC, chitosan and carboxymethyl starch* in the increasing order [62]. 500 ppm *hydroxypropyl cellulose* addition led to 90% of corrosion inhibition efficiency, whereas same amount of *gellan gum* exhibited 81% at 298 K [20]. Moreover, 0.5 g/l CMC addition into sulfuric acid revealed corrosion inhibition efficiencies of 56% and 37% for mild steel at 30°C and 60°C, respectively [56]. Moreover, increasing temperature by 30°C from room temperature led to decrease in the corrosion inhibition efficiency from 85.52% to 60.41% [43]. Hydrogen diffusion to the metal surface together with ionic mobility is accelerated by temperature rise so the blockage of adsorbed hydrogen ions is hindered and the corrosion rate is increased [43]. The decrease in the efficiency of corrosion inhibition by increasing temperature was compatible with physical adsorption of CMC on mild steel [57]. The decreasing effect of high temperature is affected by the corrosive medium. For instance, increasing temperature from 30°C to 60°C decreased the inhibition efficiency by ~ 57% in 2 M hydrochloric acid and ~ 7% in 1 M sulfuric acid when the *telfaria occidentalis* extract concentration was 50 volume percent [63]. On the other hand, the adsorption free energies were calculated, as explained before, in adsorption mechanisms and it was concluded that chemisorption mechanism overwhelms physisorption mechanism by increasing temperature from 298 K to 328 K leading to an increase in the inhibition efficiency of *gellan gum* from 81% to 85% [20]. The inhibition effect of *salvia aucheri mesatlantica* (SAMO) for steel corrosion in 0.5 M sulfuric acid was increased by increasing the temperature and the concentration of the inhibitor [23]. In addition, 0.5 g/l *gum Arabic* employed to protect steel from corrosion in 1 M sulfuric acid led to an increased inhibition efficiency from 40.1% to 52.1% due to the increase in temperature from 30°C to 60°C which is supported by chemisorption [64]. Moreover, the inhibition efficiency of 100 mg/l *tobacco rob* extract addition into artificial seawater was dramatically increased from 28% to 84% due to the increase in temperature from 25°C to 60°C that implies the chemical adsorption [65]. Similarly, chemical adsorption was the mechanism for the inhibition of mild steel in HCl solution by addition of *chitosan* because the inhibition efficiency increased up to 96% at 60°C [66]. However, decrease in the inhibition efficiency to 93% was observed upon further increase in the temperature, 70°C [66]. It may stem from the deterioration of the adsorption-desorption equilibrium towards desorption of the inhibitor [66].

7. Corrosion Prevention on Metals Other than Steel

Metals other than steel can also be protected from corrosion via natural inhibitors. Aluminum, copper, zinc and nickel were studied at different mediums. Physical adsorption process was introduced for aluminum corrosion and it was also concluded that *gum arabic* has a better effect on aluminum than steel to inhibit corrosion although the chemical adsorption was valid for the corrosion of steel [67]. The plant materials like *tobacco, lignin* and *black pepper* impeded the corrosion of aluminum in hydrochloric acid at amount of 77.5%, 67.2% and 65.8%, respectively [54]. The effects of *carboxy methyl chitosan and CMC* on the protection of aluminum corrosion in 0.5 M HCl solution were compared and higher inhibition efficiency was achieved by carboxy methyl chitosan [68]. The *acidity* of the corrosive environments has great importance on the corrosion inhibition efficiency, as in the case of steel corrosion protection. Nickel and zinc behavior was observed, besides carbon steel, in acidic, neutral and alkaline mediums including *lawsonia* which is the aqueous extract of the leaves of henna [69]. For example, the inhibition efficiency increased by increasing the acidity of the corroding medium for both carbon steel and nickel that can be attributed to the formation of complex compounds on metal surface [69]. On the other hand, the lowest inhibition efficiency was achieved for the protection of zinc in acidic medium as a result of the low stability of zinc complexes [69]. The *extracted raw material* of the natural corrosion inhibitor and the *temperature* of the medium are also effective parameters for protecting the metals in corrosive medium. *Exudate gums* extracted from the plants of *pachylobus edulis* (PE) [52, 70] and *raphia hookeri* (RH) [70] were used to protect aluminum corrosion in hydrochloric acid. Aluminum corrosion in 0.1 M hydrochloric acid was hindered by exudate gum from PE with the efficiencies of 38% and 41% at 60°C and 30°C, respectively [52]. On the other hand, the inhibition efficiency of exudate gum was increased from 41% to 56% when the extracted plant was changed from PE to RH at 30°C [70]. In addition, 0.5 g/l *ficus benjamina gum* addition into the 0.1 M sulfuric acid protected aluminum from corrosion at an efficiency level of 83.74% via chemical adsorption mechanism [71]. The *concentration* of the inhibitor in the corrosive environment also has an effect on the corrosion inhibition of metals. For instance, the effects of *rutin* and *quercetin* presented in many plants were studied on corrosion of aluminum in 3% NaCl [72]. *Rutin* is the stored version (mainly as glycosides) of quercetin in the plants [73]. The degree of protection against corrosion of aluminum in 3% NaCl was increased from 67% to 96% when *rutin* concentration was decreased from 10^{-4} to 10^{-5} mol/dm^3 [72]. This can be explained by the effect of pH values of the corrosive environments. The stability of the complex protective film formed on the aluminum surface by *rutin* and quercetin decreased at higher pH values correspond to NaCl solutions including more amount of *rutin* and quercetin [72]. Moreover, the inhibition efficiency of the *natural honey* in 0.5 M sodium chloride for copper was decreased after several days although the efficiency was increased by increasing the inhibitor concentration [74]. The decrease in the efficiency was attributed to the fungi formation since natural honey is known as an ideal medium for the growth of fungi [74]. The effects of different types of honey (oak honey, coniferous honey, winter savory honey, alder buckthorn honey and carob tree honey) on the corrosion of 5052 aluminum alloy in 0.5 M sodium chloride were studied [75]. It was concluded that the highest inhibition efficiency, 85.65%, was achieved by addition of 1200 ppm *oak honey* [75].

8. Conclusions

There are several techniques like cathodic protection and barrier preservation to protect steel and other metals from serious disadvantages of corrosion. Cathodic protection requires a system for potential accession and coatings covered before the corrosion exposure are susceptible to scrape and not economical. Therefore, including inhibitors in the corrosive medium is an important alternative for corrosion protection. In this regard, natural inhibitors have been attracting attention due to the age of increasing environmental concern about waste, pollution and toxicity. Several studies are summarized to emphasize the significant effect of adding natural inhibitors in the corrosive medium on decreasing the amount of corrosion. In addition, the efficiency of corrosion inhibition is related to the type and the concentration of the corrosion inhibitor, acidity and the temperature of the corrosive medium.

References

[1] Jacobson, G. 2016. Corrosion Measures of Prevention, Application, and Economics of Corrosion Technologies Study, NACE Int. Houston, USA. impact.nace.org/documents/Nace-International-Report.pdf, accessed 22.10.2019.
[2] Zheludkevich, M. L., D. G. Shchukin, K. A. Yasakau, H. Möhwald and M. G. S. Ferreira. 2007. Anticorrosion coatings with self-healing effect based on nanocontainers impregnated with corrosion inhibitor. Chem. Mater. 19: 402–411.
[3] Mansfeld, F., M. W. Kendig and S. Tsai. 1982. Evaluation of corrosion behaviour of coated metals with AC impedance measurements. Nat. Assoc. Corros. Eng. 38: 478–485.
[4] Yfantis, A., I. Paloumpa, D. Schmeißer and D. Yfantis. 2002. Novel corrosion-resistant films for Mg alloys. Surf. Coatings Technol. 151–152: 400–404.
[5] Shi, S. C. and C. C. Su. 2016. Corrosion inhibition of high speed steel by biopolymer HPMC derivatives. Materials (Basel) 9: 1–8.
[6] Kharmachi, I., L. Dhouibi, P. Berçot, M. Rezrazi and B. Lakard. 2017. Electrodeposition behavior, physicochemical properties and corrosion resistance of Ni–Co coating modified by gelatin additive. Prot. Met. Phys. Chem. Surfaces 53: 1059–1069.

[7] Raja, P. B., M. Ismail, S. Ghoreishiamiri, J. Mirza, M. C. Ismail, S. Kakooei and A. A. Rahim. 2016. Reviews on corrosion inhibitors: a short view. Chem. Eng. Commun. 203: 1145–1156.

[8] Chigondo, M. and F. Chigondo. 2016. Recent natural corrosion inhibitors for mild steel: an overview. J. Chem. 2016: 1–7.

[9] Satapathy, A. K., G. Gunasekaran, S. C. Sahoo, K. Amit and P. V. Rodrigues. 2009. Corrosion inhibition by Justicia gendarussa plant extract in hydrochloric acid solution. Corros. Sci. 51: 2848–2856.

[10] Taghavikish, M., N. K. Dutta and N. Roy Choudhury. 2017. Emerging corrosion inhibitors for interfacial coating. Coatings 7: 217.

[11] Wicks, Z. W. Jr., F. N. Jones, S. P. Pappas and D. A. Wicks. 2007. *In*: Organic Coatings—Science and Technology 3rd Ed., John Wiley & Sons, USA.

[12] Thomas, S., P. M. Visakh and A. P. Mathew. 2013. Advances in natural polymers: composites and nanocomposites. pp. 1–424. *In*: Adv. Struct. Mater., Springer-Verlag Berlin Heidelberg: https://doi.org/10.1007/978-3-642-14673-2.

[13] Schweitzer, P. A. 2007. Corrosion of Linings and Coatings: Cathodic and Inhibitor Protection and Corrosion Monitoring. 2nd Ed., CRC Press (Taylor & Francis Group), 1–551.

[14] Raja, P. B. and M. G. Sethuraman. 2008. Natural products as corrosion inhibitor for metals in corrosive media—A review. Mater. Lett. 62: 113–116.

[15] Stupnišek-Lisac, E., A. Gazivoda and M. Madžarac. 2002. Evaluation of non-toxic corrosion inhibitors for copper in sulphuric acid. Electrochim. Acta 47: 4189–4194.

[16] Finšgar, M. and I. Milošev. 2010. Inhibition of copper corrosion by 1,2,3-benzotriazole: A review. Corros. Sci. 52: 2737–2749.

[17] Hegazy, M. A., M. Abdallah, M. K. Awad and M. Rezk. 2014. Three novel di-quaternary ammonium salts as corrosion inhibitors for API X65 steel pipeline in acidic solution. Part I: Experimental results. Corros. Sci. 81: 54–64.

[18] Umoren, S. A. and U. M. Eduok. 2016. Application of carbohydrate polymers as corrosion inhibitors for metal substrates in different media: A review. Carbohydr. Polym. 140: 314–341.

[19] Bereket, G., C. Öretir and A. Yurt. 2001. Quantum mechanical calculations on some 4-methyl-5-substituted imidazole derivatives as acidic corrosion inhibitor for zinc. J. Mol. Struct. 571: 139–145.

[20] Rajeswari, V., D. Kesavan, M. Gopiraman and P. Viswanathamurthi. 2013. Physicochemical studies of glucose, gellan gum, and hydroxypropyl cellulose—Inhibition of cast iron corrosion. Carbohydr. Polym. 95: 288–294.

[21] Yang, L., H. Zhang, T. Tan and A. U. Rahman. 2009. Thermodynamic and NMR investigations on the adsorption mechanism of puerarin with oligo-β-cyclodextrin-coupled polystyrene-based matrix. J. Chem. Technol. Biotechnol. 84: 611–617.

[22] Ajayi, O. O., O. A. Omotosho, K. O. Ajanaku and B. O. Olawore. 2011. Degredation study of aluminum alloy in 2 M hydrochloric acid in the presence of chromolaena odorata. J. Eng. Appl. Sci. 6: 10–17.

[23] Znini, M., L. Majidi, A. Bouyanzer, J. Paolini, J. M. Desjobert, J. Costa and B. Hammouti. 2012. Essential oil of Salvia aucheri mesatlantica as a green inhibitor for the corrosion of steel in 0.5 M H_2SO_4. Arab. J. Chem. 5: 467–474.

[24] Bouklah, M., N. Benchat, A. Aouniti, B. Hammouti, M. Benkaddour, M. Lagrenée, H. Vezin and F. Bentiss. 2004. Effect of the substitution of an oxygen atom by sulphur in a pyridazinic molecule towards inhibition of corrosion of steel in 0.5 M H_2SO_4 medium. Prog. Org. Coatings 51: 118–124.

[25] Bahadori, A. 2014. Cathodic Corrosion Protection Systems: A Guide for Oil and Gas Industries. Elsevier, Oxford. https://books.google.com/books?id=7mVzAwAAQBAJ.

[26] Cruz, J., R. Martínez, J. Genesca and E. García-Ochoa. 2004. Experimental and theoretical study of 1-(2-ethylamino)-2-methylimidazoline as an inhibitor of carbon steel corrosion in acid media. J. Electroanal. Chem. 566: 111–121.

[27] Oguzie, E. E., Y. Li and F. H. Wang. 2007. Corrosion inhibition and adsorption behavior of methionine on mild steel in sulfuric acid and synergistic effect of iodide ion. J. Colloid Interface Sci. 310: 90–98.

[28] Emregül, K. C., A. Abdülkadir Akay and O. Atakol. 2005. The corrosion inhibition of steel with Schiff base compounds in 2 M HCl. Mater. Chem. Phys. 93: 325–329.

[29] Mansfeld, F. 1973. Tafel slopes and corrosion rates from polarization resistance measurements. Corrosion 29: 397–402.

[30] Fernandez-Solis, C. D., A. Vimalanandan, A. Altin, J. S. Mondragon-Ochoa, K. Kreth, P. Keil and A. Erbe. 2016. Fundamentals of electrochemistry, corrosion and corrosion protection. pp. 29–70. *In*: Lang, P. R. and Y. Liu (eds.). Fundam. Electrochem. Corros. Corros. Prot., Springer, Heidelberg/Germany: https://doi.org/10.1007/b98790.

[31] Jayaraman, A., J. C. Earthman and T. K. Wood. 1997. Corrosion inhibition by aerobic biofilms on SAE1018 steel. Appl. Microbiol. Biotechnol. 47: 62–68.

[32] Peter, A., I. B. Obot and S. K. Sharma. 2015. Use of natural gums as green corrosion inhibitors: an overview. Int. J. Ind. Chem. 6: 153–164.

[33] Zarrouk, A., H. Zarrok, R. Salghi, B. Hammouti, F. Bentiss, R. Touir and M. Bouachrine. 2013. Evaluation of N-containing organic compound as corrosion inhibitor for carbon steel in phosphoric acid. J. Mater. Environ. Sci. 4: 177–192.

[34] Bouklah, M., B. Hammouti, M. Lagrenée and F. Bentiss. 2006. Thermodynamic properties of 2,5-bis(4-methoxyphenyl)-1,3,4-oxadiazole as a corrosion inhibitor for mild steel in normal sulfuric acid medium. Corros. Sci. 48: 2831–2842.

[35] Maayta, A. K. and N. A. F. Al-Rawashdeh. 2004. Inhibition of acidic corrosion of pure aluminum by some organic compounds. Corros. Sci. 46: 1129–1140.

[36] Li, W. H., Q. He, S. T. Zhang, C. L. Pei and B. R. Hou. 2008. Some new triazole derivatives as inhibitors for mild steel corrosion in acidic medium. J. Appl. Electrochem. 38: 289–295.

[37] Lagren, M. 2002. Study of the mechanism and inhibiting 4H-1, 2, 4-triazole on mild steel corrosion in acidic media. Corros. Sci. 44: 573–588.

[38] Bahrami, M. J., S. M. A. Hosseini and P. Pilvar. 2010. Experimental and theoretical investigation of organic compounds as inhibitors for mild steel corrosion in sulfuric acid medium. Corros. Sci. 52: 2793–2803.

[39] Zarrok, H., S. S. Al-Deyab, A. Zarrouk, R. Salghi, B. Hammouti, H. Oudda, M. Bouachrine and F. Bentiss. 2012. Thermodynamic characterisation and density functional theory investigation of 1, 1',5, 5'-Tetramethyl-1H, 1'H-3, 3-Bipyrazole as corrosion inhibitor of C38 steel corrosion in HCl. Int. J. Electrochem. Sci. 7: 4047–4063.

[40] Waanders, F. B., S. W. Vorster and A. J. Geldenhuys. 2002. Biopolymer corrosion inhibition of mild steel: electrochemical/mössbauer results. Hyperfine Interact. 139–140: 133–139.

[41] Arthur, D. E., A. Jonathan, P. O. Ameh and C. Anya. 2013. A review on the assessment of polymeric materials used as corrosion inhibitor of metals and alloys. Int. J. Ind. Chem. 4: 2.

[42] Chigondo, M. and F. Chigondo. 2016. Recent natural corrosion inhibitors for mild steel: an overview. J. Chem.

[43] Hussein, M. H. M., M. F. El-Hady, H. A. H. Shehata, M. A. Hegazy and H. H. H. Hefni. 2013. Preparation of some eco-friendly corrosion inhibitors having antibacterial activity from sea food waste. J. Surfactants Deterg. 16: 233–242.

[44] Chetouani, A., B. Hammouti and M. Benkaddour. 2004. Corrosion inhibition of iron in hydrochloric acid solution by jojoba oil. Pigment Resin Technol. 33: 26–31.

[45] Elmsellem, H., A. Aouniti, M. H. Youssoufi, H. Bendaha, T. Ben hadda, A. Chetouani, I. Warad and B. Hammouti. 2013. Caffeine as a corrosion inhibitor of mild steel in hydrochloric acid. Phys. Chem. News. 70: 84–90.

[46] Aravamudhan, A., D. M. Ramos, A. A. Nada and S. G. Kumbar. 2014. Chapter 4—Natural polymers: polysaccharides and their derivatives for biomedical applications. pp. 67–89. *In*: Kumbar, S. G., C. T. Laurencin and M. Deng (eds.). Nat. Synth. Biomed. Polym.

[47] Finkenstadt, V. L., C. B. Bucur, G. L. Côté and K. O. Evans. 2017. Bacterial exopolysaccharides for corrosion resistance on low carbon steel. J. Appl. Polym. Sci. 134: 1–7.

[48] Li, M. M., Q. J. Xu, J. Han, H. Yun and Y. L. Min. 2015. Inhibition action and adsorption behavior of green inhibitor Sodium carboxymethyl cellulose on copper. Int. J. Electrochem. Sci. 10: 9028–9041.

[49] Rosliza, R. and W. B. Wan Nik. 2010. Improvement of corrosion resistance of AA6061 alloy by tapioca starch in seawater. Curr. Appl. Phys. 10: 221–229.

[50] Bello, M., N. Ochoa, V. Balsamo, F. López-Carrasquero, S. Coll, A. Monsalve and G. González. 2010. Modified cassava starches as corrosion inhibitors of carbon steel: An electrochemical and morphological approach. Carbohydr. Polym. 82: 561–568.

[51] Mobin, M., M. A. Khan and M. Parveen. 2011. Inhibition of mild steel corrosion in acidic medium using starch and surfactants additives. J. Appl. Polym. Sci. 121: 1558–1565.

[52] Umoren, S. A., I. B. Obot, E. E. Ebenso. 2008. Corrosion inhibition of aluminium using exudate gum from Pachylobus edulis in the presence of halide ions in HCl. E-Journal Chem. 5: 355–364.

[53] Abdallah, M. 2004. Guar gum as corrosion inhibitor for carbon steel in sulfuric acid solutions. Port. Electrochim. Acta 22: 161–175.

[54] Srivastav, K. and P. Srivastava. 1981. Studies on plant materials as corrosion inhibitors. Br. Corros. J. 16: 221–223.

[55] Martinez, S. 2001. Inhibitory mechanism of low-carbon steel corrosion by mimosa tannin in sulphuric acid solutions. J. Appl. Electrochem. 31: 973–978.

[56] Umoren, S. A., M. M. Solomon, I. I. Udosoro and A. P. Udoh. 2010. Synergistic and antagonistic effects between halide ions and carboxymethyl cellulose for the corrosion inhibition of mild steel in sulphuric acid solution. Cellulose 17: 635–648.

[57] Solomon, M. M., S. A. Umoren, I. I. Udosoro and A. P. Udoh. 2010. Inhibitive and adsorption behaviour of carboxymethyl cellulose on mild steel corrosion in sulphuric acid solution. Corros. Sci. 52: 1317–1325.

[58] Verbeken, D., S. Dierckx and K. Dewettinck. 2003. Exudate gums: Occurrence, production, and applications. Appl. Microbiol. Biotechnol. 63: 10–21.

[59] Umoren, S. A. and U. F. Ekanem. 2010. Inhibition of mild steel corrosion in H_2SO_4 using exudate gum from Pachylobus edulis and synergistic potassium halide additives. Chem. Eng. Commun. 197: 1339–1356.

[60] Eddy, N. O., P. Ameh, C. E. Gimba and E. E. Ebenso. 2011. GCMS studies on Anogessus leocarpus (Al) gum and their corrosion inhibition potential for mild steel in 0.1 M HCl. Int. J. Electrochem. Sci. 6: 5815–5829.

[61] El-Etre, A. Y. and M. Abdallah. 2000. Natural honey as corrosion inhibitor for metals and alloys. II. C-steel in high saline water. Corros. Sci. 42: 731–738.

[62] Saleah, A. O. and A. H. Basta. 2008. Evaluation of some organic-based biopolymers as green inhibitors for calcium sulfate scales. Environmentalist 28: 421–428.

[63] Oguzie, E. E. 2005. Inhibition of acid corrosion of mild steel by *Telfaria occidentalis* extract. Pigment Resin Technol. 34: 321–326.

[64] Umoren, S. A., O. Ogbobe, I. O. Igwe and E. E. Ebenso. 2008. Inhibition of mild steel corrosion in acidic medium using synthetic and naturally occurring polymers and synergistic halide additives. Corros. Sci. 50: 1998–2006.

[65] Wang, H., M. Gao, Y. Guo, Y. Yang and R. Hu. 2016. A natural extract of tobacco rob as scale and corrosion inhibitor in artificial seawater. Desalination 398: 198–207.

[66] Umoren, S. A., M. J. Banera, T. Alonso-Garcia, C. A. Gervasi and M. V. Mirífico. 2013. Inhibition of mild steel corrosion in HCl solution using chitosan. Cellulose 20: 2529–2545.

[67] Umoren, S. A. 2008. Inhibition of aluminium and mild steel corrosion in acidic medium using Gum Arabic. Cellulose 15: 751–761.

[68] Abdallah, M., I. Zaafarany, A. Fawzy, M. A. Radwan and E. Abdfattah. 2013. Inhibition of aluminum corrosion in hydrochloric acid by cellulose and chitosan. J. Am. Sci. 99: 580–586.

[69] El-Etre, A. Y., M. Abdallah and Z. E. El-Tantawy. 2005. Corrosion inhibition of some metals using lawsonia extract. Corros. Sci. 47: 385–395.

[70] Umoren, S. A., I. B. Obot, E. E. Ebenso and P. C. Okafor. 2007. Eco-friendly inhibitors from naturally occurring exudate gums for aluminium corrosion inhibition in acidic medium, port. Electrochim. Acta 26: 267–282.

[71] Eddy, N. O., P. O. Ameh and A. Ibrahim. 2015. Physicochemical characterization and corrosion inhibition potential of Ficus benjamina (FB) gum for aluminum in 0.1 M HCl. Walailak J. Sci. Technol. 12: 1121–1136.

[72] Berkovic, K., S. Kovac and J. Vorkapic-Furac. 2004. Natural compounds as environmentally friendly corrosion inhibitors of aluminium. Acta Aliment. 33: 237–247.

[73] Manach, C., C. Morand, C. Demigné, O. Texier, F. Régérat and C. Rémésy. 1997. Bioavailability of rutin and quercetin in rats. FEBS Lett. 409: 12–16.

[74] El-Etre, A. Y. 1998. Natural honey as corrosion inhibitor for metals and alloys. I. Copper in neutral aqueous solution. Corros. Sci. 40: 1845–1850.

[75] Gudic, S., L. Vrsalovic, M. Kliškic, I. Jerkovic, A. Radonic and M. Zekic. 2016. Corrosion inhibition of AA 5052 aluminium alloy in Nacl solution by different types of honey. Int. J. Electrochem. Sci. 11: 998–1011.

Section 2

Corrosion Protection Using Graphene and Other Smart Coatings

CHAPTER 6

Carbon Nanoallotropes-Based Anticorrosive Coatings

Recent Advances and Future Perspectives

Khaled M. Amin[1,2,]* and *Hatem M.A. Amin*[3,4,]*

1. Introduction

Corrosion is a crucial and global problem that has huge costs. Corrosion can be defined as the deterioration of a material due its reaction with its environment. Rust is the most obvious signal of corrosion. Among the different methods applied to combat corrosion, the use of coatings and corrosion inhibitors are the most common. These coatings form a sort of protective and resistant layers. Nevertheless, the choice of the corrosion protection measures should suit the substrate and its environment. Thus, an appropriate method should be chosen which ensures the most efficient protection in a particular environment. In addition, the application of various sealings or coatings (metals, non-metals, organic inhibitors, biopolymers, primers or others) can change the surface composition and add more weight to the structure. Therefore, individual measures that specify the necessary, safe and economically feasible methods of protection should be taken. An important factor in the selection of a coating is the cost, where anticorrosive coatings should only be used if they are economically favourable than an alternative corrosion-resistant material.

Today's advances in the improvement of the properties of materials such as graphene and carbon nanotubes have expanded the range of applications of various types of coatings, allowing them to adapt to the growing industry and consumers demands. Generally, carbon and its allotropes have been investigated for corrosion protection as an alternative to the toxic and hazardous inorganic and organic inhibitors [1]. Consequently, carbon-based materials offer an eco-friendlier

---
[1] Department of Materials Science, Technische Universität Darmstadt, 64287 Darmstadt, Germany.
[2] Department of Polymer Chemistry, Atomic Energy Authority, 11787 Cairo, Egypt.
[3] Department of Chemistry, Faculty of Science, Cairo University, 12613 Giza, Egypt.
[4] Faculty of Chemistry and Biochemistry, Ruhr University Bochum, 44801 Bochum, Germany.
* Corresponding authors: amin@ma.tu-darmstadt.de; hatem@pc.uni-bonn.de

solution as traditional coatings contain substances that harm aquatic life. Since its introduction in 2004, graphene has revolutionized the field of coatings [1]. In industrial applications, graphene has been used as an ideal additive for coatings [1].

Carbon and its allotropes can be classified according to their geometry/dimension into the following categories [2], as shown in Fig. 1:

i) ***0-D materials:*** such as carbon dots and fullerene

ii) ***1-D materials:*** such as carbon nanotubes and carbon nanowires

iii) ***2-D materials:*** such as graphene and graphene oxide

iv) ***3-D materials:*** such as graphite and diamond

The interest in these materials originates from their interesting properties including [3]: (i) cost parity; (ii) abundancy and availability; (iii) being environmentally friendly; (iv) simple synthesis; (v) lightness, robustness and excellent mechanical stability (especially for graphene); (vi) thermal stability; (vii) increased durability and (viii) biocompatibility. Due to the above-mentioned characteristics of most carbon allotropes, they have a huge market. For instance, in 2017 the global graphene market size was estimated to be 43 million $ and is expected to increase by 38% in 2025 [4]. Figure 2 shows the graphene market size by product in USA [4].

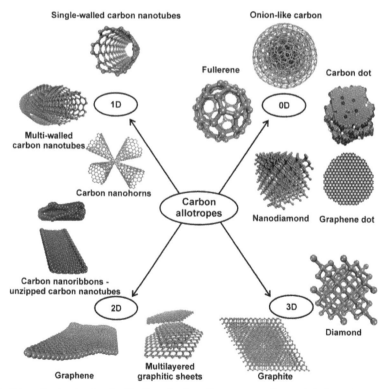

Fig. 1. Schematic showing the classification of carbon allotropes according to their dimension. Reprinted with permission from ref. [2], Copyright 2015 American Chemical Society.

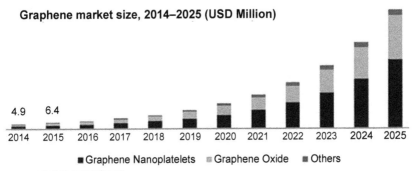

Graphene market size, 2014–2025 (USD Million)

4.9   6.4

2014 2015 2016 2017 2018 2019 2020 2021 2022 2023 2024 2025

■ Graphene Nanoplatelets ▪ Graphene Oxide ■ Others

Source: www.grandviewresearch.com

Fig. 2. Graph showing the graphene market size by product in USA, 2014–2025 (Reprinted from ref. [4]).

Carbon-based materials have been investigated for corrosion prevention and showed promising inhibition efficiency [5–7]. To further improve the inhibition ability of carbon allomorphs, several approaches, such as doping with nitrogen or sulphur, surface functionalization, incorporation of metals or combination with polymers in composites, have been reported [8–10]. It is recognized that carbon and its allotropes specially graphene hinder corrosion via a defensive mechanism, so-called barrier protection. In this mechanism, the graphene, for example, in the coating acts as a shielding layer, preventing air, salt and moisture from coming into contact with the underlying metal or alloy.

In this chapter, recent advances on the application of carbon nano-allotropes as coatings for corrosion protection of different metal substrates are reviewed. The carbon materials are categorised according to their dimension and examples of efficient coatings from each category and their performances are presented. Insights into the mechanism of inhibition are discussed. Lastly, some drawbacks, challenges and future perspectives in this field are addressed.

2. Dimension-Based Classification of Carbon Allotropes as Anticorrosive Coatings

2.1 Zero-Dimensional: Carbon Dots- and Fullerene-Based Coatings

Carbon dots (CDs) have received ever-increasing attention due to their superior properties including low toxicity, high water solubility, unique photoluminescence and biocompatibility allowing them to be used in a wide range of potential applications [11, 12]. Structure of functionalized CDs offers both multiple bonds and electronegative atoms (such as O, S and N). Moreover, they show low cytotoxicity. Consequently, these CDs can work as eco-friendly corrosion inhibitors for many metals in different corrosive environments [13–15]. CDs as interface inhibitors have three modes of action: blocking active sites, the geometric blocking and electro-catalytic effect [16].

Many researchers are focusing on the effect of doping on the enhancement of anticorrosion protection of carbon dots. Cen et al. [17] have reported the use of S and N co-doped carbon dots (N, S-CDs) for the first time as corrosion inhibitors for carbon steel in CO_2-saturated 3.5% NaCl solution. N, S-CDs showed enhanced protection and the inhibition efficiency increased as N, S-CDs contents increased (up to 93% at 50 mg/L). However, lower concentrations showed a decrease in the corrosion current compared to that of the blank condition which can be attributed to the ability of these materials to be adsorbed on the steel surface through their functional groups and N, S-CDs nanoparticles forming a hydrophilic barrier film of about 40 nm over the steel surface. Contact angle measurements confirmed the strong hydrophobicity of this barrier film. Upon the addition of N, S-CDs, the angle increased from 43° to 92° compared to 61° of thiourea under the same conditions. Hence, this carbon-based hydrophobic film offers an effective protection and good isolation of the metal surface in the corrosive media. Table 1 summarizes the changes in weight loss, corrosion rate and inhibition efficiency upon changing the concentration of CDs from 0 to 50 mg/L at 50°C for 12 hours. The decrease in corrosion current density (i_{corr}) with the addition of the doped CDs confirmed the hinderance of the charge transfer process.

Another study investigated the synthesis and corrosion behaviour of functionalized carbon dots (FCDs) in 1 M HCl solution [18]. FCDs were synthesized by conjugation of citric acid carbon dots (CA-CDs) and imidazole. The electrochemical characterizations showed that the FCDs can effectively reduce the corrosion rate of Q235 steel in 1 M HCl medium with an inhibition efficiency up to 90% in case of FCDs content of 100 mg/L. According to the Langmuir adsorption model, the excellent protection property could be attributed, as discussed above, to the good coverage of the formed adsorption film of FCDs on the metal surface. Furthermore, Nyquist plots obtained from electrochemical impedance spectroscopy (EIS) showed that the diameter of capacitive reactance arc increased upon the addition of the different compositions (imidazole, CA-CDs and FCDs), confirming the corrosion inhibition of steel in presence of these solutions [19]. This diameter increased with the increase of the inhibitor concentration in case of imidazole and FCDs, while it was stable in case of CA-CDs which means that the barrier adsorbed film over the steel metal becomes denser with the increase of imidazole and FCDs concentrations. Inhibition efficiency calculated from EIS date revealed significant difference in efficiency between the different inhibitors with the increase of inhibitor concentration. Q235 steel revealed inhibition efficiency of 39.1% in case of

Table 1. Changes in weight loss, corrosion rate and inhibition efficiency of carbon steel at 50°C with different additions of N, S-CDs for 12 hours (Reprinted with permission from ref. [17], Copyright 2019 Elsevier).

C_{CDs} (mg L^{-1})	ΔW (g)	Corrosion rate (g m^{-2} h^{-1})	η (%)
0	0.0188 ± 0.0015	1.15 ± 0.092	–
10	0.0061 ± 0.0003	0.37 ± 0.019	67.8
20	0.0033 ± 0.0002	0.20 ± 0.012	82.6
30	0.0026 ± 0.0001	0.16 ± 0.008	86.0
40	0.0016 ± 0.0001	0.10 ± 0.005	91.3
50	0.0013 ± 0.0001	0.08 ± 0.006	93.0

200 mg/L CA-CDs, while the efficiency significantly increased in case of using the same concentration of imidazole and FCDs (82.7% and 93.4%, respectively). These results were consistent with the inhibition efficiencies obtained from Tafel plots.

To investigate the presence of localized corrosion of steel, the study applied the Scanning Vibrating Electrode Technique (SVET) (Fig. 3). As the steel electrode in pure HCl solution was exposed to severe corrosion, it showed the highest positive current. On the other hand, a significant decrease in the current was observed in case of imidazole and FCDs solutions and the highest decrease in current was observed for FCDs. This observation confirms that the corrosion inhibition is induced by the FCDs. Moreover, as shown in Fig. 3, current density decreased with the increase in inhibitor concentration and immersion time.

Fullerenes, which are considered as 0-D analogues of benzene, are carbon-enriched molecules generally of ellipsoidal and spherical morphology [20]. Fullerene has a lubricating behaviour which paves the way for their use in tribology and industrial applications. This behaviour arises from their strong intermolecular

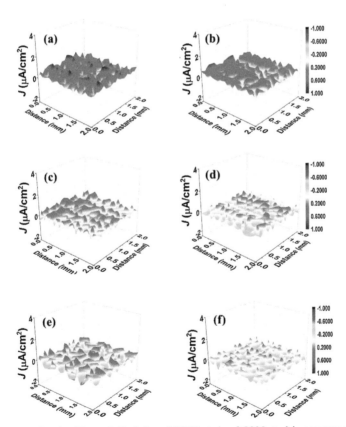

Fig. 3. Scanning Vibrating Electrode Technique (SVET) study of Q235 steel in presence of different solutions and immersion times: (a) 0%–12 h; (b) CA-CDs-200 mg/L-12 h; (c) imidazole-200 mg/L-12 h; (d) FCDs-200 mg/L-12 h; (e) FCDs 100 mg/L-12 h; (f) FCDs-200 mg/L-24 h (Reprinted with permission from ref. [18], Copyright 2019 Elsevier).

bonding and spherical morphology [21]. There are limited studies investigating the performance of fullerene and its composites in corrosion protection field compared to other carbonaceous materials like graphene and carbon nanotubes [22–25].

One of the most recently published studies reported the incorporation of Oxidized Fullerene (OF) nanoparticles into a hybrid coating of silane (TEOS+GPTMS) [26]. The researchers used low and moderate concentrations of OF in the range of 25–100 mg/L. Fourier-transform infrared spectroscopy (FT-IR) and Raman measurements confirmed the successful interaction between the sol-gel coating and OF nanoparticles. Dynamic light scattering (DLS) measurement of OF (concentration of 100 mg/L) dispersed in the coating showed a mean size of 1798 nm for the OF nanoparticles with good dispersion which was also confirmed by Transmission Electron Microscope (TEM) micrographs. The average roughness (R_a) values, derived from Atomic Force Microscope (AFM) micrographs, of the blank coating gradually decreased by inclusion of different concentrations of OF up to 100 mg/L. This decrease in the roughness can be related to the formation of dense layer as a result of the interaction between the coating and OF nanoparticles [27]. However, the roughness increased by inclusion of higher concentrations of OF (200 and 500 mg/L) which may be attributed to OF nanoparticles agglomeration and defects formation on the metal surface. EIS measurements confirmed the improvement in corrosion resistance upon incorporation of increased concentrations of OF nanoparticles (25–100 mg/L); however, further increase in OF concentration (200 and 500 mg/L) resulted in a decrease in the corrosion resistance. The micromorphology revealed a decrease in the corrosion damages by inclusion of 25 mg/L OF nanoparticles. These damages were completely removed in case of 50 mg/L OF nanoparticles due to the formation of a defect free isolating layer, which in turn prevents the diffusion of the corrosive solution to the metal substrate.

2.2 One-Dimensional: CNT-Based Coatings

Carbon nanotubes (CNTs) are basically graphene sheets rolled up to form seamless tubes. They are considered as superior reinforcing materials for different composites because of their excellent mechanical properties such as stiffness, modulus and strength compared to other existing materials [28]. CNTs have showed electrochemical behaviour different from the other carbon allotropes [29–31]. Many previous studies have reported the enhancement in corrosion resistance provided by CNTs for different metals as a result of the modification of surface by filling and blocking the crevices and holes [32–36]. However, CNTs have a great tendency to aggregation in aqueous solutions because of the high length-to-diameter ratio which represents the major drawback facing the application of CNTs in corrosion protection field as it leads to deterioration of the mechanical and chemical properties of the coatings [37]. To overcome this problem, several studies have investigated different approaches such as addition of different surfactants to the solution, ball-milling process and functionalization to prevent the aggregation and improve the dispersibility [38–40].

Several studies investigated the industrial applications of CNTs in corrosion protection field. Among them, Goyal et al. [41] studied the protection efficiency of different CNT-enhanced Cr_3C_2-20NiCr matrices on ASME-SA213-T-22 steel. The effect of concentration variation of CNTs (1 wt% and 2 wt%) prepared using high-velocity oxyfuel (HVOF) thermal spraying method on the corrosion behaviour and mechanical properties at 900°C in super-heater zone of actual coal-fired boiler of a thermal power plant was investigated. The coating's thickness was in the range of 270–280 μm. The roughness measurements showed a decrease in roughness with increase in CNTs concentration; on the other hand, the hardness increased with the increase in CNTs concentration. The bare electrode showed high rate of corrosion because of the formed porous Fe_2O_3 layer, while Cr_3C_2–20NiCr coatings decreased the corrosion rate by 77.3% which is attributed to the formation of spinels of Ni and Co onto the surface. Figure 4 shows the differences in corrosion rates and mass gain between bare T22 steel and the steel coated with different compositions. Cr_3C_2-20NiCr-2 wt% CNT demonstrated the highest corrosion protection efficiency due to the high ability to block the pores and to prevent the electrolyte from reaching the metal surface.

Another study addressed the enhanced hydrophobicity and corrosion protection induced by the incorporation of different concentrations of multi-walled carbon nanotube (MWCNT) reinforced $Ni/MoSe_2$ composite coatings on Ni substrates in 3.5% NaCl [40]. The modified composites showed enhanced hydrophobicity compared to the blank coating which is attributed to the addition of a second phase reinforcement. Moreover, the modified coatings showed more compact morphology compared to the blank Ni coating. As shown in Fig. 5a, the coatings with 0.5 and 1 g/L MWCNT loadings displayed lower i_{corr} values, confirming nobler behaviour of the two coatings. The protection efficiency calculated from the polarization measurements proved the enhanced protection of the MWCNT reinforced coatings, $Ni-MoSe_2$-0.1 g/L, and it revealed the maximum efficiency (94.5%). In consistence with the previous observation, Nyquist plot of $Ni-MoSe_2$-0.1 g/L MWCNT coating showed the largest diameter of the semicircle arc followed by that of $Ni-MoSe_2$-0.5 g/L MWCNT coating. Moreover, the fitting data revealed higher charge-transfer resistance (R_{ct}) for both the coatings, confirming higher protection efficiency compared to the other coatings (Fig. 5b).

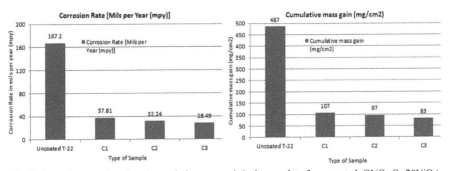

Fig. 4. Corrosion rates (mpy) and cumulative mass gain/unit area plots for uncoated, C1(Cr_3C_2-20NiCr), C2 (Cr_3C_2-20NiCr-1%CNT), and C3 (Cr_3C_2-20NiCr-2%CNT) coated T22 samples after exposure to actual boiler environment at 900°C (Reprinted with permission from ref. [41], Copyright 2019 Taylor & Francis).

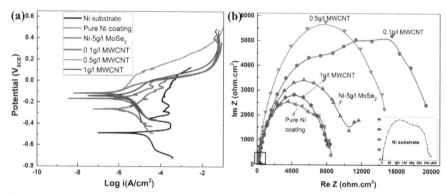

Fig. 5. (a) Tafel plot and (b) Nyquist plot of bare and coated Ni substrate in 3.5 wt% NaCl solution (Reprinted with permission from ref. [40], Copyright 2019 Elsevier).

Combination of CNTs with conducting polymers in coatings can significantly improve the corrosion protection performance of such coatings. This could be due to a synergistic effect between both components in their mixture. Conducting polymers have already been used as alternative candidates in corrosion protection due to their low cost, ease of synthesis and good chemical stability [42–45]. The excellent corrosion protection ability of conducting polymers such as polyaniline (PANI) mainly arises from passivation of catalysis and electron transmission on metal surfaces, which in turn relies on the redox activity of the PANI itself [46]. In this context, researchers are working on enhancing the protection ability of such materials through the formation of nanocomposites to control the structure and morphology and to increase the redox activity.

One of the most recent works, conducted by Rui et al. [47], investigated an effective strategy to prepare PANI/CNT core-shell nanocomposites with excellent redox activity in both acidic and neutral medium via oxidative polymerization of aniline on the surface of calcium carbonate and CNT nanocomposites. The as-prepared PANI/CNT nanocomposites were further dispersed into acrylate-amino (AA) resin to form anticorrosive coatings for mild steel in 3.5 wt% NaCl solution. TEM investigation confirmed the enhanced uniform dispersity of PANI/CNT (4:1) in the AA coating compared with single nanomaterials because of its larger surface area [24, 48]. The roughness measurements of AA, 1 wt% PANI-AA and 1 wt% PANI/CNT-AA coatings before immersion in the electrolyte showed no clear difference in hydrophobicity, indicating the homogeneous distribution of the prepared nanofillers in AA matrix. The corrosion protection efficiency, calculated from the potentiodynamic polarization curves of 1 wt% PANI/CNT-AA coating (99.996%), was higher than the corresponding value of 1 wt% PANI-AA coating. The elucidated protection times in 3.5 wt% NaCl electrolyte at 65°C for the AA coating incorporated with 0.5, 1 and 3 wt% of PANI/CNT nanocomposite were 76, 76 and 55 days, respectively, which indicate the stability of PANI/CNT dispersion in AA matrix, which in turn is reflected in the passivation of the metal surface. Nyquist plots of EIS measured after extended immersion times (2 and 5 days) revealed a gradually increasing radius of the single-capacity arc, demonstrating the formation of a dense oxide film on the metal surface, while the Nyquist plot showed a double-capacitance arc when corrosion protection failed, which indicates the complete penetration of the corrosive medium into the coating. By studying the mechanism of action of PANI/ CNT nanocomposites as corrosion protection coatings, it was revealed that this layer works as an efficient physical barrier to corrosive ions. The high dispersion stability of PANI/CNT can provide an enhanced diffusion resistance to water and oxygen in the corrosive medium because the nanoscale additives like CNTs give better physical barrier properties compared to microscale additives [49–52].

2.3 Two-Dimensional: Graphene- and Graphene Oxide-Based Coatings

Graphite derivatives such as graphene and graphene oxide are considered as promising anti-corrosion coatings for metals due to their superior properties including lightweight, impermeability, inertness, wear resistance and mechanical stability. The chemical inertness is introduced as one of the strongest justifications for their use as anti-corrosion coatings [53]. Graphite-based materials have proven enhanced potential as fillers for polymeric matrices that can be efficiently used as a barrier or protective layer for most of metals against corrosion. The effective distribution of such materials into the polymer matrix can be achieved by different ways including surface functionalization and selection of polymer with suitable functional groups that can interact with graphite derivatives to form some kind of physical or chemical bonding [54].

The effect of exfoliated graphite loading on the galvanic corrosion protection of the aluminium alloy AA7075 using unfunctionalized exfoliated graphite/polyetherimide (UFG/PEI) nanocomposites have been studied [55]. The electrochemical characterization revealed two different modes of corrosion protection according to the concentration of exfoliated graphite, as shown in Fig. 6. At low concentrations of exfoliated graphite, which does not exceed the percolation threshold, the applied coatings showed long-term corrosion protection for the metal and this can be attributed to the inclusion of the exfoliated graphite into the polymeric matrix. On the other hand, at high concentrations of exfoliated graphite, which exceed the percolation threshold, the corrosion protection becomes very difficult and the galvanic corrosion of the aluminium alloy starts in propagation due to the large differential in redox potentials between the metal substrate itself and the exfoliated graphite layers as indicated in Fig. 6. The electronic transport properties of graphite-based materials mainly depend on the thickness of the forming layers, while the barrier properties are mainly controlled by reducing the ion transport which is preserved even for the multilayer graphene and the functionalized graphene derivatives. When the compatibility between exfoliated graphite and the polymeric matrix is low, the layers tend to agglomerate and restack again forming phase segregated domains. These domains do not provide effective tortoise pathways, and as a result this effect reduces the ion transport rate and enhances the corrosion protection. This study reported that the exfoliated graphite loadings over the threshold value (10 and 17 wt% in PEI) cause differential redox potentials, which in turn increases the corrosion rate of electropositive metals.

 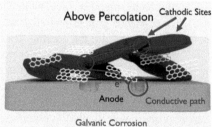

Fig. 6. Modes of action of exfoliated graphite-based coatings based on the loading: at low exfoliated graphite content up to 5 wt% UFG (Left scheme) and at high exfoliated graphite content up to 10 wt% UFG or higher (Right scheme) (Adapted with permission from ref. [55], Copyright 2019 American Chemical Society).

Inclusion of different metals in the graphene matrix attracts a great attention as an effective route to enhance the corrosion protection efficiency of metal-based coatings due to the increased surface area offered by the graphene matrix and the better distribution of the metal particles in the coatings.

The effect of graphene loading in Ni/graphene nanocomposites on the mechanical, morphological properties and corrosion resistance behaviour of steel substrates in 3.5 wt% NaCl solution was reported [56]. The electrochemical co-deposition of nickel and graphene nanosheets to form the Ni/graphene coatings was achieved and the graphene content was controlled by changing the graphene concentration in the electrolyte solution. The surface roughness of the different coatings has been investigated using SEM and the data indicated that the roughness of the graphene-containing coatings increased compared to graphene-free nickel coatings. Furthermore, AFM measurements and SEM micrographs indicated the increased coarseness of the coatings as the graphene concentration in the deposition electrolyte increases (from 100 to 400 mg/L), as shown in Fig. 7.

The graphene content also affected the microhardness of the coatings which is significantly increased by increasing the graphene content in the coatings, which in turn decreases the grain size and reduces the dislocation sliding in the nickel matrix and hence increases the hardness of the matrix [57]. Tafel plots showed increased corrosion potentials (E_{corr}) and decreased corrosion currents upon the increase in graphene content in the coatings. The higher carbon content not only enhances the mechanical strength of the coatings but also fills the gaps and blocks the voids and cracks which are formed during the electrodeposition of the Ni film over the steel due to the small size of graphene sheets leading to enhanced corrosion protection of the graphene-containing coatings [57].

In an alternative approach, gamma irradiation was used to prepare graphene (GIG) via reduction of GO and it was then applied as a coating for protection of AISI 316 stainless steel in 3.5% NaCl solution [7]. Field Emission Scanning Electron Microscope (FE-SEM) and TEM displayed the flake-like structure of GIG sheets with a lateral dimension of 1 μm. GIG was further modified through the inclusion of chitosan (CS) into GIG matrix to form GIG/CS composite coating. Potentiodynamic polarization and EIS measurements proved high protection efficiency offered by

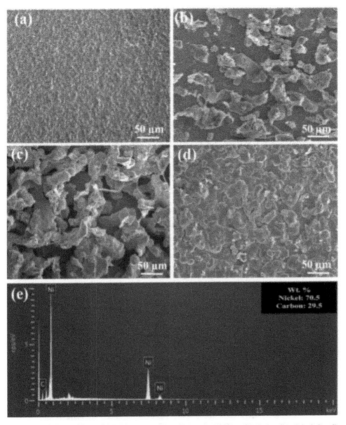

Fig. 7. SEM micrographs of Ni-graphene coatings with (a) 0 g/L, (b) 0.1 g/L, (c) 0.2 g/L and (d) 0.4 g/L graphene in the deposition bath. EDS spectrum of Ni-graphene coating with 0.4 g/L graphene in depositing bath (e) (Reprinted with permission from ref. [56], Copyright 2018 Elsevier).

these coatings. i_{corr} in case of GIG and GIG/CS coatings was one order of magnitude higher than that of bare AISI 316 steel. The effect of temperature (between 298 and 328 K) on the potential-current response of different coatings was studied and the data indicated that the increased temperature has a limited effect on i_{corr} values of GIG and GIG/CS coated surfaces compared to that of the bare surface. Moreover, the apparent activation energy of bare AISI 316, GIG-coated and GIG/CS-coated were 2.92, 5.23 and 6.76 kJ mol^{-1}, respectively, confirming the high protection efficiency of graphene-based coatings. This study also reported the EIS measurements of bare and coated steel surfaces after different immersion times in the electrolyte solution and the data indicated higher resistance from GIG and GIG/CS coatings than from the bare steel and thus confirmed the stability of the coatings after different immersion times in NaCl solution. The values of R_{ct} were higher for GIG and GIG/CS coated stainless steel compared to the bare surface which means that these coatings impart a protective layer for stainless steel against pitting corrosion in sodium chloride-containing electrolytes. In addition, surface studies revealed the exposure of bare steel surface to pitting corrosion after immersion in 3.5% NaCl solution, while examination of coated surfaces after removal of GIG and GIG/CS coatings showed intact steel surface which was free of pitting spots (Fig. 8).

In a further work conducted by the same research group [5], the corrosion behaviour of GIG coatings on AISI 316 steel was investigated in another media, 0.5 M sulfuric acid solution. Electrochemical measurements indicated the enhanced protection efficiency of GIG (up to 76%) which had been further enhanced upon the incorporation of an isoquinoline-based organic inhibitor (PDTI) with GIG (up to 95%). Investigation of surface morphology of bare AISI 316 steel and coatings of GIG after immersion for 72 hours in 0.5 M sulfuric acid solution exhibited a relatively compact structure where microscopic voids were not identified.

Graphite derivatives can enhance the corrosion protection through another approach: they can play a key role in improving the adhesion of the different coatings, specially for coatings which suffer from brittleness and weak resistance to crack corrosion [23, 58, 59]. A recent study reported the effect of reinforcing the epoxy composite coatings by different concentrations of reduced graphene oxide (RGO) on the corrosion behaviour of N80 substrates in 10 wt% NaCl [60]. The

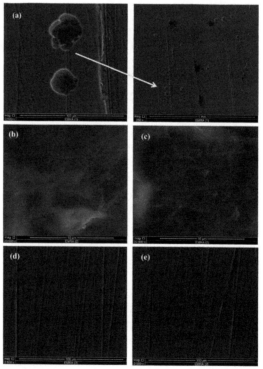

Fig. 8. FE-SEM of (a) bare AISI 316 steel after immersion in 3.5% NaCl and subjected to Tafel test (low and high magnifications), (b) and (c) AISI 316 steel surfaces with GIG and GIG/CS before immersion in 3.5% NaCl, respectively, (d) and (e) the AISI 316 steel surfaces with GIG and GIG/CS films removed, respectively, and after immersion in 3.5% NaCl for 2 hours and subjected to Tafel test (Reproduced with permission from ref. [7], Copyright 2015 The Royal Society of Chemistry).

scratch test results showed that the spalling width in the epoxy coating decreased upon the incorporation of different concentrations of graphene as indicated in Fig. 9, which confirms the improvement in the coatings adhesion. Polarization curves and the calculated kinetic parameters indicated enhanced protection efficiency for the RGO/epoxy coatings compared to epoxy coatings. The epoxy coating containing 1 wt% RGO showed the lowest i_{corr} (0.137 μA/cm^2) and the highest E_{corr} (0.343 V) compared to the other concentrations. Fitting data of EIS measurements suggested that both the coating resistance and the charge transfer resistance R_{ct} increase with the increase in RGO concentration up to 1 wt% RGO, then they decrease with further increase in RGO concentration up to 4 wt%, confirming the results obtained from Tafel plots. This also indicates that 1 wt% RGO is the optimum concentration for modification of epoxy coatings that offers the maximum protection efficiency under the investigated conditions.

In another study, the distribution of graphene oxide (GO) in solvent-based epoxy coatings for protection of mild steel in NaCl solution was explored [61]. The study followed two approaches for preparation of the coatings to determine the optimum procedure for preparation of stable coatings with excellent resistance. The first route involved the addition of GO directly to epoxy resin followed by the addition of polyamide hardener (R+GO) +H, while the second route involved the addition of GO to the hardener before the addition of epoxy resin R+ (H+GO). Morphology examinations indicated that the coatings prepared using R+ (H+GO) method displayed better exfoliation of GO sheets due to the higher adhesion of the sheets to the polymer matrix and no pull-out of GO from the fracture was observed compared to the coatings prepared using R+ (H+GO) due to the strong covalent bonding between GO and polymer matrix [62]. Polarization curves showed that GO-modified coatings have different behaviour compared to the values of pure epoxy coatings. (R+GO) +H coating exhibited more negative E_{corr} and lower i_{corr} than the neat epoxy coating. In contrast, R+ (H+GO) coating revealed more positive E_{corr} and significant decrease in i_{corr} than the values of pure epoxy coating, confirming a better

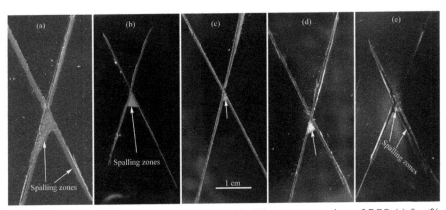

Fig. 9. Macro-morphology of the epoxy coatings with different concentrations of RGO (a) 0 wt%, (b) 0.5 wt%, (c) 1.0 wt%, (d) 2.0 wt%, and (e) 4.0 wt% after measuring adhesion via the scratch method (Reprinted with permission from ref. [60], Copyright 2019 Elsevier).

dispersion and enhanced barrier properties of the nanocomposite coatings prepared via R+ (H+GO) compared to those prepared via (R+GO) +H scenario. This is due to lower viscosity of polyamide hardener compared to that of epoxy resin. Moreover, anodic and cathodic Tafel slopes were changed after the inclusion of GO into the coatings, suggesting the synergistic effect between GO and the coating.

In this chapter, we focused the discussion on 0D, 1D and 2D materials rather than the 3D materials since their properties render them more promising materials in coatings field.

3. Concluding Remarks, Drawbacks and Future Perspectives

The extraordinary characteristics of carbon-based films, specially graphene and its composites, including mechanical stability, thermal conductivity, flexibility and easy surface functionalization render them eco-friendly alternatives to the harmful inorganic and organic corrosion inhibitors. The excellent barrier properties lead to an improved inhibition efficiency by effectively blocking the penetrating species [63]. Importantly, graphite and carbon derivatives are abundant, cost-effective and light materials. Although graphene, fullerene, carbon dots, etc. have been investigated in the last decade as anticorrosive materials, and some of them have already been included in industrial coatings, there are still some challenges and possible perspectives that have to be addressed in future research:

i) *Upscaling*: The production of graphene and carbon-based nanomaterials can be scaled up by using advanced large engineering technologies. However, it is worth mentioning that the optimum parameters for large scale synthesis of carbon materials and testing their inhibition efficiency on industrial level could be different from the optimized conditions on the lab scale. Accordingly, we believe that developing efficient coatings and upscaling the application of such coatings should be done in parallel.

ii) *Safety-related issues*: Due to the nano-size of these materials, specially carbon fibres and carbon nanotubes, these materials could penetrate into the cells and thus can be considered toxic [64]. However, the use of carbon nanomaterials with specific characteristics could avoid or at least minimize their risks. Nevertheless, the safety risks of carbon nanomaterials are entirely dependent on the type of the carbon nanostructure, surface chemistry, shape, size and how it is applied [64]. Therefore, generalization of the nanotoxicity of carbon nanomaterials is misleading and this issue should be handled case by case. Research efforts should be done to explore the long-term health impact and the fate of these materials in the environment, and importantly how these particles enter the cell and cause the damage.

iii) *Quality of graphene*: The conductivity of graphene and graphite films vary significantly depending on the surface defects, orientation, synthesis procedure and purity. This causes various reaction rates at the surface [65]. Therefore, challenges related to quality control of graphene and carbon nanotubes materials and its influential role in the inhibition efficiency still remain unclear. Moreover, combining single layer of graphene with graphite or polymers changes the conductivity, and this is an effective procedure to improve the inhibition ability. In addition, aggregation of nanostructures is still problematic and thus current procedures to mitigate this effect have to be optimized.

iv) *Synergy in mixed nano-hybrids*: Carbon nano-allotropes are believed to perform better if they are combined with other anti-corrosive substances. In such hybrids, graphene layers, for example, work predominantly as the barrier and provide the mechanical stability for the coating as in the case of graphene oxide reinforced epoxy resins [63]. Thus, the development of hybrid materials made of multiple carbon nano-allotropes offer practical opportunities to produce materials with combined properties of their constituents via synergistic and cooperative effects. As a result, the hybrid carbon-based nanomaterials can find applications in fields where they were uncompetitive with classical materials. Therefore, further work on investigating more efficient and green combinations for sustainable coatings is needed.

4. Acknowledgement

H.M.A. Amin would like to thank the Faculty of Science, Cairo University for their support.

References

[1] Cui, G., Z. Bi, R. Zhang, J. Liu, X. Yu and Z. Li. 2019. A comprehensive review on graphene-based anti-corrosive coatings. Chem. Eng. J. 373: 104–121.

[2] Georgakilas, V., J. A. Perman, J. Tucek and R. Zboril. 2015. Broad family of carbon nanoallotropes: classification, chemistry, and applications of fullerenes, carbon dots, nanotubes, graphene, nanodiamonds, and combined superstructures. Chem. Rev. 115: 4744–4822.

[3] Novoselov, K. S., A. K. Geim, S. V. Morozov, D. Jiang, Y. Zhang, S. V. Dubonos, I. V. Grigorieva and A. A. Firsov. 2004. Electric field in atomically thin carbon films. Science 306: 666–669.

[4] Graphene Market Size | Industry Growth Analysis Report. 2019–2025. Grand view research, https://www.grandviewresearch.com/industry-analysis/graphene-industry/request/rs2 (accessed February 9, 2020).

[5] Galal, A., K. M. Amin, N. F. Atta and H. A. Abd El-Rehim. 2017. Protective ability of graphene prepared by γ-irradiation and impregnated with organic inhibitor applied on AISI 316 stainless steel. J. Alloys Compd. 695: 638–647.

[6] Hong, M. -S., Y. Park, T. Kim, K. Kim and J. -G. Kim. 2020. Polydopamine/carbon nanotube nanocomposite coating for corrosion resistance. J. Mater. 6: 158–166.

[7] Atta, N. F., K. M. Amin, H. A. Abd El-Rehim and A. Galal. 2015. Graphene prepared by gamma irradiation for corrosion protection of stainless steel 316 in chloride containing electrolytes. RSC Adv. 5: 71627–71636.

[8] Prasannakumar, R. S., V. I. Chukwuike, K. Bhakyaraj, S. Mohan and R. C. Barik. 2020. Electrochemical and hydrodynamic flow characterization of corrosion protection persistence of nickel/multiwalled carbon nanotubes composite coating. Appl. Surf. Sci. 507: 145073.

[9] Chilkoor, G., R. Sarder, J. Islam, K. E. ArunKumar, I. Ratnayake, S. Star, B. K. Jasthi, G. Sereda, N. Koratkar, M. Meyyappan and V. Gadhamshetty. 2020. Maleic anhydride-functionalized graphene nanofillers render epoxy coatings highly resistant to corrosion and microbial attack. Carbon N. Y. 159: 586–597.

[10] Tian, Y., Y. Xie, F. Dai, H. Huang, L. Zhong and X. Zhang. 2020. Ammonium-grafted graphene oxide for enhanced corrosion resistance of waterborne epoxy coatings. Surf. Coatings Technol. 383: 125227.

[11] Cui, M., S. Ren, Q. Xue, H. Zhao and L. Wang. 2017. Carbon dots as new eco-friendly and effective corrosion inhibitor. J. Alloys Compd. 726: 680–692.

[12] Song, Y., C. Zhu, J. Song, H. Li, D. Du and Y. Lin. 2017. Drug-derived bright and color-tunable N-doped carbon dots for cell imaging and sensitive detection of Fe^{3+} in living cells. ACS Appl. Mater. Interfaces 9: 7399–7405.

[13] Pourhashem, S., E. Ghasemy, A. Rashidi and M. R. Vaezi. 2018. Corrosion protection properties of novel epoxy nanocomposite coatings containing silane functionalized graphene quantum dots. J. Alloys Compd. 731: 1112–1118.

[14] Cui, M., S. Ren, H. Zhao, L. Wang and Q. Xue. 2018. Novel nitrogen doped carbon dots for corrosion inhibition of carbon steel in 1 M HCl solution. Appl. Surf. Sci. 443: 145–156.

[15] Singh, A., K. R. Ansari, A. Kumar, W. Liu, C. Songsong and Y. Lin. 2017. Electrochemical, surface and quantum chemical studies of novel imidazole derivatives as corrosion inhibitors for J55 steel in sweet corrosive environment. J. Alloys Compd. 712: 121–133.

[16] Cao, C. 1996. On electrochemical techniques for interface inhibitor research. Corros. Sci. 38: 2073–2082.

[17] Cen, H., Z. Chen and X. Guo. 2019. N, S co-doped carbon dots as effective corrosion inhibitor for carbon steel in CO_2-saturated 3.5% NaCl solution. J. Taiwan Inst. Chem. Eng. 99: 224–238.

[18] Ye, Y., D. Yang and H. Chen. 2019. A green and effective corrosion inhibitor of functionalized carbon dots. J. Mater. Sci. Technol. 35: 2243–2253.

[19] Dou, W., J. Liu, W. Cai, D. Wang, R. Jia, S. Chen and T. Gu. 2019. Electrochemical investigation of increased carbon steel corrosion via extracellular electron transfer by a sulfate reducing bacterium under carbon source starvation. Corros. Sci. 150: 258–267.

[20] Zhang, Z., L. Wei, X. Qin and Y. Li. 2015. Carbon nanomaterials for photovoltaic process. Nano Energy 15: 490–522.

[21] Zhai, W., N. Srikanth, L. B. Kong and K. Zhou. 2017. Carbon nanomaterials in tribology. Carbon N. Y. 119: 150–171.

[22] Sittner, F., B. Enders and W. Ensinger. 2004. Electrochemical determination of corrosion protection ability of fullerene thin films treated by radiofrequency plasma. Thin Solid Films 459: 233–236.

[23] Liu, D., W. Zhao, S. Liu, Q. Cen and Q. Xue. 2016. Comparative tribological and corrosion resistance properties of epoxy composite coatings reinforced with functionalized fullerene C60 and graphene. Surf. Coatings Technol. 286: 354–364.

[24] Madhan Kumar, A. and Z. M. Gasem. 2015. Effect of functionalization of carbon nanotubes on mechanical and electrochemical behavior of polyaniline nanocomposite coatings. Surf. Coatings Technol. 276: 416–423.

[25] Sittner, F., B. Enders, H. Jungclas and W. Ensinger. 2002. Corrosion properties of ion beam modified fullerene thin films on iron substrates. Surf. Coatings Technol. 158–159: 368–372.

[26] Samadianfard, R., D. Seifzadeh, A. Habibi-Yangjeh and Y. Jafari-Tarzanagh. 2020. Oxidized fullerene/sol-gel nanocomposite for corrosion protection of AM60B magnesium alloy. Surf. Coatings Technol. 385: 125400.

[27] Ashrafi-Shahri, S. M., F. Ravari and D. Seifzadeh. 2019. Smart organic/inorganic sol-gel nanocomposite containing functionalized mesoporous silica for corrosion protection. Prog. Org. Coatings 133: 44–54.

[28] Shahedi Asl, M., I. Farahbakhsh and B. Nayebi. 2016. Characteristics of multi-walled carbon nanotube toughened ZrB2-SiC ceramic composite prepared by hot pressing. Ceram. Int. 42: 1950–1958.

[29] Pierson, H. O. 1993. Handbook of Carbon, Graphite, Diamonds and Fullerenes: Processing, Properties and Applications (Materials Science and Process Technology). Noyes Publications.

[30] Shao, Y., G. Yin, J. Zhang and Y. Gao. 2006. Comparative investigation of the resistance to electrochemical oxidation of carbon black and carbon nanotubes in aqueous sulfuric acid solution. Electrochim. Acta 51: 5853–5857.

[31] Yazdani, S., F. Mahboubi, R. Tima and O. Sharifahmadian. 2019. Effect of carbon nanotube concentration on the corrosion behavior of electroless Ni-B-CNT coating. J. Mater. Eng. Perform. 28: 3446–3459.

[32] Bijalwan, P., K. K. Pandey, B. Mukherjee, A. Islam, A. Pathak, M. Dutta and A. K. Keshri. 2019. Tailoring the bimodal zone in plasma sprayed CNT reinforced YSZ coating and its impact on mechanical and tribological properties. Surf. Coatings Technol. 377: 124870.

[33] Alishahi, M., S. M. Monirvaghefi, A. Saatchia and S. M. Hosseini. 2012. The effect of carbon nanotubes on the corrosion and tribological behavior of electroless Ni-P-CNT composite coating. Appl. Surf. Sci. 258: 2439–2446.

[34] Arora, S., N. Kumari and C. Srivastava. 2019. Microstructure and corrosion behaviour of NiCo-Carbon nanotube composite coatings. J. Alloys Compd. 801: 449–459.

[35] Wang, X., Z. Li, W. Zhan, J. Tu, X. Zuo, X. Deng and B. Gui. 2019. Preparation and corrosion resistance of titanium-zirconium/nickel-coated carbon nanotubes chemical nano-composite conversion coatings. Anti-Corrosion Methods Mater. 66: 343–351.

[36] Cubides, Y. and H. Castaneda. 2016. Corrosion protection mechanisms of carbon nanotube and zinc-rich epoxy primers on carbon steel in simulated concrete pore solutions in the presence of chloride ions. Corros. Sci. 109: 145–161.

[37] Zhou, M., Y. Mai, H. Ling, F. Chen, W. Lian and X. Jie. 2018. Electrodeposition of CNTs/copper composite coatings with enhanced tribological performance from a low concentration CNTs colloidal solution. Mater. Res. Bull. 97: 537–543.

[38] Albaaji, A. J., E. G. Castle, M. J. Reece, J. P. Hall and S. L. Evans. 2017. Effect of ball-milling time on mechanical and magnetic properties of carbon nanotube reinforced FeCo alloy composites. Mater. Des. 122: 296–306.

[39] Chiang, Y. C., W. H. Lin and Y. C. Chang. 2011. The influence of treatment duration on multi-walled carbon nanotubes functionalized by H_2SO_4/HNO_3 oxidation. Appl. Surf. Sci. 257: 2401–2410.

[40] Maharana, H. S., P. K. Katiyar and K. Mondal. 2019. Structure dependent super-hydrophobic and corrosion resistant behavior of electrodeposited Ni-MoSe$_2$-MWCNT coating. Appl. Surf. Sci. 478: 26–37.

[41] Goyal, K. and R. Goyal. 2019. Improving hot corrosion resistance of Cr_3C_2–20NiCr coatings with CNT reinforcements. Surf. Eng. 0: 1–10.

[42] Shchukina, E., D. Grigoriev, T. Sviridova and D. Shchukin. 2017. Comparative study of the effect of halloysite nanocontainers on autonomic corrosion protection of polyepoxy coatings on steel by salt-spray tests. Prog. Org. Coatings 108: 84–89.

[43] Sababi, M., J. Pan, P. E. Augustsson, P. E. Sundell and P. M. Claesson. 2014. Influence of polyaniline and ceria nanoparticle additives on corrosion protection of a UV-cure coating on carbon steel. Corros. Sci. 84: 189–197.

[44] Zaarei, D., A. A. Sarabi, F. Sharif, M. M. Gudarzi and S. M. Kassiriha. 2012. A new approach to using submicron emeraldine-base polyaniline in corrosion-resistant epoxy coatings. J. Coatings Technol. Res. 9: 47–57.

[45] Zhang, Y., Y. Shao, X. Liu, C. Shi, Y. Wang, G. Meng, X. Zeng and Y. Yang. 2017. A study on corrosion protection of different polyaniline coatings for mild steel. Prog. Org. Coatings 111: 240–247.

[46] Zhu, A., H. Wang, S. Sun and C. Zhang. 2018. The synthesis and antistatic, anticorrosive properties of polyaniline composite coating. Prog. Org. Coatings 122: 270–279.

[47] Rui, M., Y. Jiang and A. Zhu. 2020. Sub-micron calcium carbonate as a template for the preparation of dendrite-like PANI/CNT nanocomposites and its corrosion protection properties. Chem. Eng. J. 385: 123396.

[48] Deshpande, P. P., S. S. Vathare, S. T. Vagge, E. Tomšík and J. Stejskal. 2013. Conducting polyaniline/multi-wall carbon nanotubes composite paints on low carbon steel for corrosion protection: Electrochemical investigations. Chem. Pap. 67: 1072–1078.

[49] Kalendová, A., I. Sapurina, J. Stejskal and D. Veselý. 2008. Anticorrosion properties of polyaniline-coated pigments in organic coatings. Corros. Sci. 50: 3549–3560.

[50] Gupta, B., A. Rakesh, A. A. Melvin, A. C. Pandey and R. Prakash. 2014. *In-situ* synthesis of polyaniline coated montmorillonite (Mt) clay using Fe^{3+} intercalated Mt as oxidizing agent. Appl. Clay Sci. 95: 50–54.

[51] Liao, G., Q. Li and Z. Xu. 2019. The chemical modification of polyaniline with enhanced properties: A review. Prog. Org. Coatings 126: 35–43.

[52] Ramezanzadeh, B., G. Bahlakeh and M. Ramezanzadeh. 2018. Polyaniline-cerium oxide (PAni-CeO$_2$) coated graphene oxide for enhancement of epoxy coating corrosion protection performance on mild steel. Corros. Sci. 137: 111–126.

[53] Kyhl, L., S. F. Nielsen, A. G. Čabo, A. Cassidy, J. A. Miwa and L. Hornekær. 2015. Graphene as an anti-corrosion coating layer. Faraday Discuss. 180: 495–509.

[54] Dennis, R. V., V. Patil, J. L. Andrews, J. P. Aldinger, G. D. Yadav and S. Banerjee. 2015. Hybrid nanostructured coatings for corrosion protection of base metals: a sustainability perspective. Mater. Res. Express. 2: 032001.

[55] Davidson, R. D., Y. Cubides, C. Fincher, P. Stein, C. McLain, B. -X. Xu, M. Pharr, H. Castaneda and S. Banerjee. 2019. Tortuosity but not percolation: design of exfoliated graphite nanocomposite coatings for extended corrosion protection of aluminum alloys. ACS Appl. Nano Mater. 2: 3100–3116.

[56] Yasin, G., M. A. Khan, M. Arif, M. Shakeel, T. M. Hassan, W. Q. Khan, R. M. Korai, Z. Abbas and Y. Zuo. 2018. Synthesis of spheres-like Ni/graphene nanocomposite as an efficient anti-corrosive coating; effect of graphene content on its morphology and mechanical properties. J. Alloys Compd. 755: 79–88.

[57] Ren, Z., N. Meng, K. Shehzad, Y. Xu, S. Qu, B. Yu and J. K. Luo. 2015. Mechanical properties of nickel-graphene composites synthesized by electrochemical deposition. Nanotechnology 26.

[58] Monetta, T., A. Acquesta, F. Bellucci. 2015. Graphene/epoxy coating as multifunctional material for aircraft structures. Aerospace 2: 423–434.

[59] Wan, Y. J., L. X. Gong, L. C. Tang, L. Bin Wu and J. X. Jiang. 2014. Mechanical properties of epoxy composites filled with silane-functionalized graphene oxide. Compos. Part A Appl. Sci. Manuf. 64: 79–89.

[60] Zhu, L., C. Feng and Y. Cao. 2019. Corrosion behavior of epoxy composite coatings reinforced with reduced graphene oxide nanosheets in the high salinity environments. Appl. Surf. Sci. 493: 889–896.

[61] Pourhashem, S., M. R. Vaezi, A. Rashidi and M. R. Bagherzadeh. 2017. Exploring corrosion protection properties of solvent based epoxy-graphene oxide nanocomposite coatings on mild steel. Corros. Sci. 115: 78–92.

[62] Wang, X., W. Xing, P. Zhang, L. Song, H. Yang and Y. Hu. 2012. Covalent functionalization of graphene with organosilane and its use as a reinforcement in epoxy composites. Compos. Sci. Technol. 72: 737–743.

[63] Hou, W., Y. Gao, J. Wang, D. J. Blackwood and S. Teo. 2020. Recent advances and future perspectives for graphene oxide reinforced epoxy resins. Mater. Today Commun. 23: 100883.

[64] Bussy, C., H. Ali-Boucetta and K. Kostarelos. 2013. Safety considerations for graphene: lessons learnt from carbon nanotubes. Acc. Chem. Res. 46: 692–701.

[65] Kirkland, N. T., T. Schiller, N. Medhekar and N. Birbilis. 2012. Exploring graphene as a corrosion protection barrier. Corros. Sci. 56: 1–4.

CHAPTER 7

Analytical Techniques for Corrosion-Related Characterization of Graphene and Graphene-Based Nanocomposite Coatings

Saman Hosseinpour,[1,]* *Ali Davoodi,*[2,3,]* *Arash Sedighi*[2] and *Faeze Tofighi*[2]

1. Introduction

Per a simplified definition, corrosion is the tendency of a metal to convert into metal oxide deteriorating the performance of the metal surface. The corrosion rate, in addition to the type of metal, thus depends on the environmental factors, temperature, and mechanical forces involved. Moreover, the electrochemical interactions between different parts of a metallic substrate and also its shape may influence its local corrosion rate. Protective coatings are among the best ways to prevent the corrosion of metals in the industry. These coatings include metallic or alloy based organic layers, silanes, conductive polymers, oxide layers, and even thiol-based, selenol-based or carboxylic acid-based monolayers [1–5]. Nevertheless, scratching or cracking of coatings and other imperfections in protective coatings are among the phenomena that may arise from the difference between the substrates, affecting the corrosion resistance properties of the coating/metal system [2, 6]. This issue is especially common in the thick protective layers.

Graphene and its composites, as the thinnest barrier coatings with significant anti-corrosion characteristics, have recently been proved to be promising candidates against corrosion of metals [2, 7, 8]. Graphene is a 2D structure of carbon atoms with

[1] Institute of Particle Technology (LFG), Friedrich-Alexander-Universität-Erlangen-Nürnberg (FAU), Cauerstraße 4, 91058 Erlangen, Germany.
[2] Materials and Metallurgical Engineering Department, Faculty of Engineering, Ferdowsi University of Mashhad, Mashhad 91775-1111, Iran.
[3] Harvard University, John A. Paulson School of Engineering and Applied Sciences, Gordon McKay Laboratory, 29 Oxford Street, Cambridge MA 02130, USA
* Corresponding authors: hsaman@kth.se; adavoodi@seas.harvard.edu

a hexagonal crystalline structure with sp^2 bonds. Graphene's appealing properties include high surface area/mass ration (providing excellent adsorption tendency and surface reactions), electron mobility, thermal conductivity, and mechanical strength. With the highest surface area/mass ratio compared to any other material of this kind, graphene nanosheets, often in the form of nano-composites, possess a high tendency to interact with polymer or metal surfaces [1, 7]. Nevertheless, the corrosion protection efficiency that such coatings provide depends upon multiple factors, including substrate properties, defect structure, and density in the coating as well as exposure condition, and external mechanical and thermal stresses. In this respect, a thorough characterization of graphene-based coatings, often using more than one analytical tool, deems necessary. Such characterization tools should be able to interrogate the coating/metal system at different length scales from macro to micro and nano-scale.

Various characterization methods have been recently employed in studying graphene-based nanocomposite coatings. These methods include microscopic and spectroscopic tools, thermal stability assessment, and electrochemical tests [9–12]. Some of these characterization techniques are widely used in the corrosion community, whereas some others are less common, more specific, state-of-art, and sophisticated. Nevertheless, the complementary application of these techniques in a multi-analytical assessment approach often provides a deeper and more fundamental understanding of the properties and efficiency of these coating under various exposure conditions. Therefore, a well-designed and detailed strategy is required to select and use the proper set of techniques to answer a specific question regarding the structure and corrosion protection performance of graphene-based coatings in a given exposure condition. Such a strategy should be based on necessities, cost-effectiveness, availability, ease of use, and other demands. This chapter serves as a guide to help in making the right and quick decision about the most suitable techniques as well as complementary ones when it comes to the assessment of graphene-based coatings for the protection of surfaces against corrosion. Eventually, the best corrosion performances that are provided by graphene-based coatings are also provided as golden benchmarks for such coatings in industrial applications.

2. Imaging Examinations

2.1 SEM and TEM Imaging

Morphological and microstructural examinations of graphene and graphene-based nanocomposite coatings are usually performed by SEM, TEM, and AFM imaging techniques, the details of which have been described elsewhere [13, 14]. Due to the 2D structure of graphene, often, SEM and TEM investigations are performed complementarily to provide a complete picture of the microstructure of graphene-based coatings. For instance, SEM and complementary TEM images of graphene and 3, 4, 9, 10-perylene tetracarboxylic acid-graphene (PTCA-G) composite are shown in Fig. 1 [15]. SEM and TEM images of graphene (Fig. 1A and C) exhibit the typical wrinkled structure. The rough paper-like structure of graphene can be observed in this SEM image. The images for PTCA-G composite coating indicates PTCA stacks

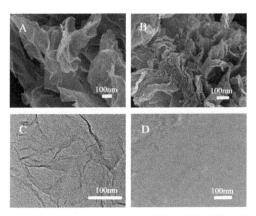

Fig. 1. SEM (A), TEM (C) images of graphene and SEM (B), TEM (D) images of PTCA-G composites [15].

in graphene with a face-to-face structure due to the π-π interactions between the sp² plane of graphene and the five-benzene core of PTCA. SEM image of PTCA-G (Fig. 1B) displays the flake-like structure, whereas the corresponding TEM image (Fig. 1D) depicts the hierarchical nanostructure with a slightly crumpled morphology [15]. This is one of the examples in which the complementary picture of the structure of a graphene-based composite coating is obtained using more than one imaging technique.

The introduction of high-resolution TEM (HRTEM) facilitated resolving even more delicate structures of graphene-based coatings. For instance, nanosheets of graphene oxide (GO) were captured at 200 kV [16], as depicted in Fig. 2A–D. Selim et al. also studied the structure of nanosheets and topology of nanocoating using field emission SEM (FESEM) at 30 kV. Nevertheless, obtaining high-resolution TEM images requires proper and relatively complicated sample preparation. Proper samples can be prepared by ultra-sonication of the GO nanosheets in ethyl

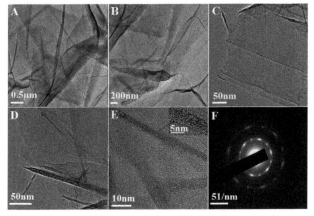

Fig. 2. (A–D) The TEM images of the as-synthesized GO nanosheets at various magnifications; (E) Layers of the prepared nano GO and (F) corresponding SAED pattern [16]

alcohol (for 15 min) followed by dropping a few drops of the mixture on a TEM grid. Furthermore, the selected area electron diffraction (SAD) method enables analyzing the electron diffraction pattern of nanosheets. The sheet-like morphology of the graphene layer accompanied by folded regions at the edges can be noted in Fig. 2E. The diffraction spots with exfoliated symmetry obtained by SAED (Fig. 2F) are consistent with a hexagonal lattice [16, 17].

The effect of polycrystalline graphene (PG) on the corrosion inhibition of the solvent-based epoxy coating was studied recently [18]. The coated substrates were broken in cryo temperature under liquid nitrogen and were coated by a thin layer of gold to avoid charge accumulation on the sample during imaging. The average size of exfoliated nano-sheets was close to 50 nm. The size of dispersed nano-fillers in the matrix directly influences the properties of nano-composite. The authors also studied the fracture surface of the epoxy/PG nano-composite coating using FE-SEM (Fig. 3). Pure epoxy coatings are often brittle and present a smooth and featureless fracture surface morphology because of their weak resistance against crack initiation and propagation. The fracture surface of nanocomposite coatings was, however, rough and river-like, as the result of intrinsic reinforcement nature of 2D PG sheets within

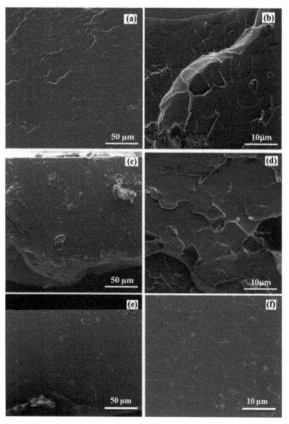

Fig. 3. FE-SEM images from cross-sections of epoxy/PG nano-composite coatings containing different amount of PG; (a, b) 0.1 wt.% PG, (c, d) 0.3 wt.% PG, and (e, f) 0.5 wt.% PG [18].

the composite coating. The fractography of nano-composite coatings also showed that the PG sheets are dispersed in polymer matrix randomly. For nano-composites coating with 0 up to 0.1 weight percentage PG, it was observed that PG sheets, with their wavy structure, are homogeneously dispersed in the polymer matrix. Nevertheless, for coatings with higher PG content, agglomeration of reinforcing particles was observed. The high magnified FE-SEM investigations revealed the existence of some gaps between PG sheets and matrix for nanocomposite coatings containing 0.3 and 0.5 weight percentage PG, which were associated with the weak adhesion at the interface between PG sheets and epoxy coating. In this situation, de-bonding occurs during the sample fracture and a river-like structure forms on the fracture surface [18].

Environmental-friendly waterborne polyurethane/graphene oxides nano-composites (WPU/GOs) were manufactured in a comprehensive study using p-tert-butyl calix [4] arene (BC4A), and sodium p-sulfonatocalix [4] arene (SC4A) modified GO nanosheets (hereafter referred to as CGO and SGO, respectively) [6]. The authors showed that the unmodified GO sheets are stacked tightly due to the high agglomeration tendency. Moreover, magnified image demonstrates that GO has a slightly smooth surface. In the case of SGO, highly roughened surface, more densely thin-layered and wrinkled morphology was obtained compared to those in the CGO coatings. The authors correlated their observation to the repulsion forces between sulfonate groups of SC4 as attached to the surface of GO nanosheets that could avoid the stacking or agglomeration of BC4A modified sheets. The higher stacked morphology of CGO in comparison with SGO was attributed to the self-assembly mechanism of BC4A macrocycles on the surface of GO [6].

As Wang et al. showed, folding of the GO sheets starts from the edges toward the base under delayed reduction, and a wrinkled structure appears after 6 hr reaction time. This observation suggests that oxygen-containing functional groups are decreased after a long reduction time. The surface of epoxy that was used as the matrix of the coating in their studies became a rough ladder-like fault structure by the addition of the reduced GO sheets [19].

The GO sheets with different aspect ratios were synthesized by Jiang et al. by controlling the amounts of oxidants and oxidation time. These GO sheets were denoted from high to low as GO-a, GO-b, and GO-c and their corresponding SEM images are shown in Fig. 4. All the GO sheets presented a lamellar morphology with high transparency, indicating that GO layers were quite thin [8, 17].

A uniform reduced graphene oxide/poly (vinyl alcohol) (RGO/PVA) composite coating was successfully fabricated on the Mg alloy surface, through a simple spin-assisted layer-by-layer assembly (SA-LBL) technology by Chu et al. [20]. The surface micrographs of RGO-PVA hybrid coatings are shown in Fig. 5 in which wrinkles and folds can be observed in all of the samples. Figure 5a shows the morphology of a drop cast-coated reduced GO film exhibiting rough wrinkles on the surface compared to the spin-coated samples. Figure 5b indicates that in SA-LBL, as a practical deposition method, the surface quality of the coating is improved by increasing the PVA weight percentage from 0 to 90. The number of surface wrinkles also decreases, and the high-density fine wrinkles convert to a coarse network-like

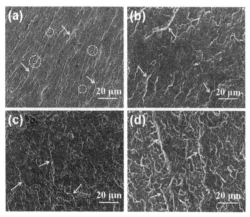

Fig. 4. SEM images of fracture surfaces for (a) pure epoxy, (b) GO-a/EP, (c) GO-b/EP, and (d) GO-c/EP coatings [8].

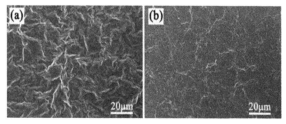

Fig. 5. SEM surface micrographs of RGO/PVA hybrid coatings: (a) drop-coated RGO, (b) pure RGO coating [20].

structure upon increasing the PVA content in the coating. The latter effect results in the formation of a smoother surface later. These phenomena were related to the reduced proportion of reduced GO with the increase of PVA in each layer [20, 21].

The cross-sectional micrographs of reduced graphene oxide coatings with and without PVA addition are provided in Figure 6. The thickness of the reduced graphene oxide coating was measured to be 3.95 µm, which decreases by increasing the weight percentage PVA. In Fig. 6a, the reduced graphene oxide sheets are stacked revealing a loose layered structure due to the weak interaction between the adjacent reduced GO sheets. Figure 6b, however, represents a well-organized and uniform morphology. As the cross-linker content in the coating is increased to 70 wt.% the cross-section becomes more continuous than coating containing lower cross-linker content (Fig. 6c) [20].

In another research, polyaniline (PANI) nanofibers-CeO_2 grafted graphene oxide nanosheets were synthesized through a layer-by-layer (LbL) assembly approach [22]. The authors used SEM to visualize that PANI covers the whole surface of GO-CeO_2 and provides different morphologies (see Fig. 7). The authors also showed that a combination of CeO_2 particles and PANI fibers with primary sizes less than 100 nm covers the GO sheets [9, 22, 23].

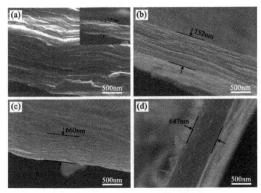

Fig. 6. The cross-section SEM micrographs of (a) pure reduced GO coating, (b) reduced GO/PVA hybrid coating (50 wt.% PVA), (c) 70 wt.% PVA, (d) 90 wt.% PVA [20].

Fig. 7. FE-SEM micrographs of GO-PANI-CeO$_2$ nanosheets [22].

In the coatings studied by Zhan et al., the Fe$_3$O$_4$ nanoparticles were uniformly coated on the surface of graphene oxide via hydrothermal route and GO-Fe$_3$O$_4$ dispersed in dopamine and secondary functional monomer (GO-Fe$_3$O$_4$ @ poly (DA+KH550)) was prepared. For TEM characterization, the authors prepared samples by placing drops of the diluted GO-Fe$_3$O$_4$ @ poly (DA+KH550) hybrid aqueous suspension onto a carbon-coated copper grid. After drying in an ambient oven, the samples were placed in the TEM chamber. Figure 8a confirms that Fe$_3$O$_4$ nanoparticles are dispersed on the surface of graphene oxides. It is evident from Fig. 8b that the co-modification of the composite coating with KH550 and dopamine improves the interfacial adhesion between nanoparticles and graphene oxide. As it is evident from Fig. 8b and c, the graphene oxide and Fe$_3$O$_4$ nanoparticles are covered with organic coating, which is the consequence of the reaction and copolymerization of KH550 and dopamine on the surface of GO-Fe$_3$O$_4$. From the high magnification TEM (Fig. 8d), the thickness of the coating was determined to be almost 25 nm. It was also observed that, nanoparticles with a size of about 20–100 nm were anchored onto the wall of the graphene foam [24].

As demonstrated by multiple examples, complementary SEM and TEM analysis allows a deeper characterization of graphene-based coatings and composite protective layers. Another equally important technique to study the surface and properties of such coatings is AFM, which will be discussed in the next section.

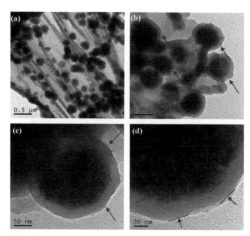

Fig. 8. TEM images of GO-Fe₃O₄@poly (KH550+DA) hybrid with different magnifications (a–d) as indicated in the magnification bar [24].

2.2 AFM Imaging

AFM has been extensively applied for evaluating the surface roughness, as well as other interfacial properties, of the composite coatings [25]. Yasin et al. evidenced the formation of single-layered graphene oxide nanosheets through AFM. The size of the GO sheets in their study was almost 5 μm, and the thickness of GO sheets was below 1.5 nm. In their studies, the sample preparation for AFM measurements was performed by depositing dilute GO suspensions on the cleaned silicon followed by a drying step at room temperature in air. As is depicted in Fig. 9, the GO sheets display different thicknesses and lateral dimensions. The corresponding height cross-sectional of AFM images (Fig. 9) revealed that the average thickness of GO-a, GO-b, and GO-c was 4.8, 4.3, and 3.7 nm, respectively. Besides, the area distribution of GO sheets was evaluated by an average of 100 sheets for each sample [8, 16, 24] revealing the largest surface area distribution for GO-a.

Reduced graphene oxide containing titanate coupling agents of two-branched dioctylpyrophosphate (T2) and three-branched dioctylpyrophosphate (T3) were used in a study by Wang et al. [26]. They used AFM imaging (as shown in Fig. 10) to reveal the surface structure of the fabricated coatings. In general, the thickness of the monolayer graphene is 0.34 nm. The authors thus measured the interlayer spacing of T2-RGO and T3-RGO [26] and found that in comparison with T2-containing reduced graphene oxide layer, the increased height of T3-RGO indicates an increased interlayer spacing between nanosheets. This was attributed to the intercalated T3 of a higher branching degree compared to T2. It is noteworthy that the nanosheet size of T2- and T3-incorporated reduced graphene oxide becomes much bigger than that in pure reduced GO, a phenomenon which is due to the incorporation of a titanate coupling agent that can protect reduced GO nanosheets [26].

Images of graphene oxide/polyurethane (GO-PU) nano-composites were obtained with an AFM operated in dynamic (also known as tapping) mode by Romani et al., as well as other researchers. The authors used the AFM topography

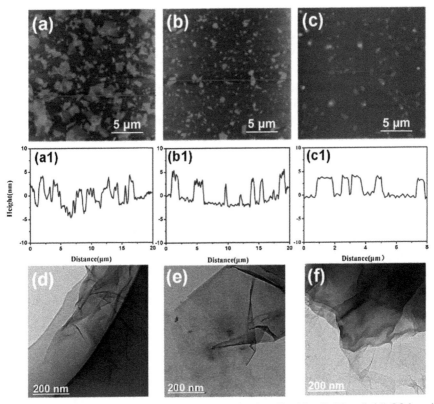

Fig. 9. AFM images and corresponding height cross-sectional profile of (a, a1) GO-a, (b, b1) GO-b, and (c, c1) GO-c. Corresponding TEM images for GO-a, GO-b, and GO-c are provided in (d), (e) and (f), respectively [8].

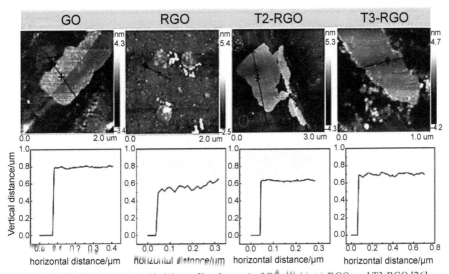

Fig. 10. AFM images (top) and height profiles (bottom) of GO, RGO, T2-RGO, and T3-RGO [26].

Table 1. Surface roughness RMS (nm) of the coating PU, GO-PU, and reduced GO-PU [11].

Coating	PU	GO-PU	RGO-PU
RMS/nm	11 ± 1	18 ± 2	17 ± 2

images to calculate the root mean square (RMS) roughness of coatings. Examples of these values are presented in Table 1. The values in this table show that the roughness of the coatings increases slightly by the incorporation of GO and reduced GO fillers [11, 12, 17].

Besides the abovementioned imaging techniques, often the information on the chemical nature of the coatings is required for understanding and predicting its corrosion resistance properties. In the following section, various spectroscopic techniques that have been utilized in the examination of graphene-based protective layers are introduced and some of their exemplary results are provided.

3. Spectroscopic Examinations

3.1 EDS Examination

As was mentioned earlier, the amount ratio of different constituents of composite coatings plays an important role in determining the characteristics of the coating, including their corrosion resistance. Therefore, various spectroscopic examinations are often applied to measure the percentage of elements in samples and composite coatings. For instance, Asgar et al. studied the purity of their fabricated composite coating and its chemical states with the help of energy dispersive spectroscopy (EDS) [27]. As was stated, GO is a composition of carbon graphene and oxygen with a mass ratio of 5.5. The EDS pattern thus can be utilized in characterizing GO containing coatings and to reveal the distribution of different components in GO nanosheets. For instance, Fig. 11 shows the EDS pattern of carbon and oxygen in the as-synthesized GO nanosheets [16].

In the nickel–graphene composite coatings that were prepared by different electrodeposition current densities, the carbon content in the composite coatings was found to increase gradually with the increasing of the current density up to a certain level. Moreover, the EDS studies revealed that graphene nanosheets also incorporate the nickel matrix, due to the reduction of nickel ions around them, as graphene

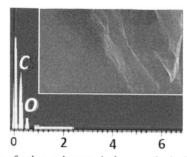

Fig. 11. EDS pattern of carbon and oxygen in the as-synthesized GO nanosheets [16].

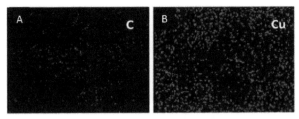

Fig. 12. EDS spectrum of Cu-10G composite, A carbon, and B, pure Cu [29].

is a more conductive compound than nickel. The limited diffusion of graphene nanosheets from the electrolyte to the cathode results, however, in reaching a steady-state by a further increase of the addition rate of graphene in the coatings [25, 28]. As can be observed in Fig. 12, the small graphene clusters can be detected based on the distribution of carbon. These clusters form due to their stacking tendency owing to the strong π-π bond as well as van der Waals force. Clusters shown in the EDS spectrum (Fig. 12) are for Cu-10G composite coating [29].

3.2 FTIR Spectroscopy

Fourier transform infrared (FTIR) spectroscopy is another common spectroscopic method that is often utilized for the characterization of the carbon-containing coatings. Despite the fact that, generally, FTIR spectrometers lack the required surface sensitivity and lateral resolution for studying the local properties of a heterogeneous structure, the systematic design of experiments facilitates unraveling the important parameters affecting the performance of the relatively complex structures. Furthermore, recent developments in IR spectroscopy, such as IR reflection absorption spectroscopy (IRRAS), polarization modulation IR (PM-IRRAS), application of IR microscopy and the introduction of near field techniques (e.g., tip enhanced-IR spectroscopy) allow reaching a reasonable lateral (and sometimes temporal) resolution, *in situ* and *ex situ*.

Chu et al., for instance, used a conventional IR spectroscopy in the reflection mode to characterize the functional groups present on the surface of graphite, graphene, and graphene oxide and examined the change of functional groups during the pre and post chemical modifications. In addition, they used IR spectroscopy to analyze the evolution of chemical bonds in all of the samples [17, 20, 30].

In Fig. 13, examples of IR spectra of graphite and graphene oxide are provided. The detailed discussion about the analysis of the FTIR spectra of GO as a corrosion inhibitor has been provided elsewhere [27]. In the case of reduced graphene oxides, the main characteristic absorption peaks in the IR spectra include –COOH and –C–O–C– vibrations at 1722 cm^{-1} and 1052 cm^{-1}, respectively. The intensity of these peaks was observed to decrease gradually with the reduction time gradually. This reduction of the peak intensities was correlated with the removal of numerous oxygen-containing functional groups during the processes of the film. GO spectra also exhibit a broad peak at 3368 cm^{-1} which has been attributed to the O-H bond stretching vibrations, likely as hydroxylated carbons (C-OH). Furthermore, the absorption frequencies corresponding to the vibrations of other functional groups

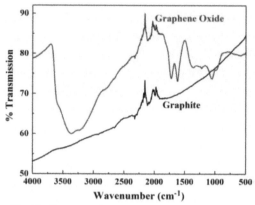

Fig. 13. IR spectra of graphite and graphene oxide [27].

include =CH and aromatic C=C at 2925 and 1624 cm⁻¹, respectively, confirming the effective synthesis of GO nanosheets covered with carboxylic and hydroxyl at the surface. In the analysis of the IR spectra, the peaks at 2325–1981 cm⁻¹ were characteristics of the diamond crystal used [19, 27, 30].

Nonahal et al. analyzed the FTIR spectrum of the functionalized GO (FGO) and associated the peaks at 3649 and 1573 cm⁻¹ to the presence of N-H bond stretching frequency and bending frequency, respectively. The sharp peak at 1506 cm⁻¹ was associated with the stretching frequency of formed C-N bond, whereas the peaks at 1711 and 1400 cm⁻¹ were correlated to C=O and C-O bonds of carbonyls of the amide bond, respectively. Other vibration modes such as C-O-C stretching frequency of epoxy and nitro groups present in the FGO were identified at 1215, 1548, and 1400 cm⁻¹, respectively [23, 31].

Addition of GO and reduced GO changes the position of carboxyl (-COOH) in epoxy/(reduced)GO composite coatings. Moreover, the –OH stretching band is sensitive to hydrogen bonding and can be shifted to a lower wavenumber, if the strength of the hydrogen bonding increases, as observed for the addition of GO in an epoxy matrix. This shift was stronger than that if reduced GO was added to the epoxy matrix, indicating more extensive hydrogen-bonding interactions between functional groups in the system (e.g., between the oxygen-containing and hydroxyl functional groups of graphene and the epoxy) [19].

In the case of FTIR spectra of amine-terminated polyamidoamine (PAMAM) and GO-PAMAM, the characteristic band of carboxyl groups of GO almost disappears in the spectrum of GO-PAMAM. This could be related to the consumption of the carboxyl groups in the course of the amidation process during surface functionalization. The intensity of the absorption peak for C=C stretching vibration, which exists in the graphene structure, decreases for GO-PAMAM because these groups often react with the modifier molecules. Moreover, for PAMAM modified GO, a new band appears in the IR spectrum which has been associated with the N-H bending vibration of amide groups (at 1560 cm⁻¹). The presence of this band is direct evidence ensuring the successful linkage of the dendrimer molecules on the GO surface [23, 31]. In another research by Yang et al., the FTIR spectra of 3, 4, 9, 10-perylene tetracarboxylic acid-graphene (PTCA-G) composite were reported [15]. The authors observed a reduction in the υ(C=C) characteristic peaks (at 1595 and 1629 cm^{-1}) due to the formation of the π-π stacking interactions and hydrophobic forces between PTCA and graphene [15].

In the modification of GO nanosheets by BC4A and SC4A, Mohammadi et al. utilized FTIR spectroscopy to demonstrate that the BC4A macrocycles successfully bond to the surface of GO nanosheets [6]. The appearance of adsorption bands at 2959 cm^{-1} (C-H asymmetric stretching), 2867 cm^{-1} (C-H symmetric stretching), and 1115 cm^{-1} (S=O asymmetric stretching) provided evidence that GO was modified successfully by SC4A macrocycles, as well. The decrease in adsorption intensity of the epoxide groups after modification of GO with C4As in their research was justified by the reaction of C4As with epoxide groups through a ring-opening reaction [6].

3.3 Raman Spectroscopy

Similar to the FTIR spectroscopy, Raman spectroscopy has also been extensively utilized by the researchers to study the properties of the graphene and graphite containing coatings and corrosion inhibiting layers and to study the faith of the coatings upon exposure to corrosive media. Specifically, Raman spectroscopy is considered as a powerful analytical technique to identify the defects and disorder degree of graphene-based structures and graphene nanosheet [11, 12]. The D band arising from the defects and disorder structure of carbon in graphene plaques as well as the G band originating from the first-order scattering of the E_{2g} vibration mode and the in-plane vibration of ordered sp^2-bonded carbon atoms are the main peaks that are used as a quality measure for the assessment of graphene layers in the coating. In this respect, the defects and disorder degree of each sample can be quantified based on the peak intensity ratio of the D band to the G band (I_D/I_G) which reports on the sp^3/sp^2 carbon ratio and is considered as a scale of the extent of defects [6]. Mohammadi et al. also observed that the degree of defect and disorder in CGO and SGO is slightly more than GO because of the presence of C4As macrocycle on the surface of nanosheets [6]. In fact, G and D Raman peaks of graphene and graphene oxide (GO) signify sp^2 hybridization and the degree of disorder due to the defects induced on the sp^2 hybridized hexagonal sheet of the carbon. A small peak intensity ratio of I_D/I_G suggests the graphitization in the as-grown graphene structure. The frequency range of G and D Raman shifts is in the range of 1500–1600 cm^{-1} and 1300–1400 cm^{-1}, respectively. In the case of GO, the intensity of D peak is more significant than that for G peak, demonstrating a more disordered structure. In another word, the structural defects within the carbon lattice of grapheme result in a stronger D peak. The peak from 2650–2700 cm^{-1} is called second order-disorder mode for carbon nanomaterial and is another important peak which is utilized in studying graphene/graphite containing coatings [30], and denotes that the as-grown graphene sheets are of multilayer type or not.

Typical Raman spectroscopy analysis is often performed using the excitation wavelength of 532 nm laser, the spot size of 500 nm, and the exposure time of 10 s. The laser power can be up to 0.2 mW to avoid the laser-induced sample damage. In the Raman spectrum shown in Fig. 14, the I_D/I_G of graphene is about 1.42 owing to

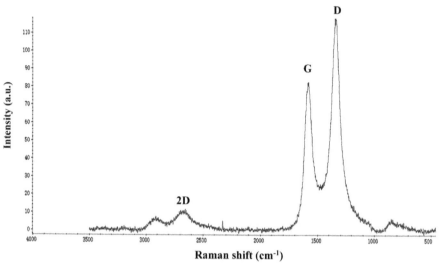

Fig. 14. Raman spectra of graphene nanosheet and graphite [33].

the presence of unrestored defects that stay after the elimination of large amounts of oxygen-containing functional groups. This I_D/I_G ratio value is consistent with most chemical reduction reports [32, 33]. Furthermore, the average crystallite size of the sp^2 domains in typical GO nanosheets can be estimated based on the ratio I_D/I_G, according to the Tuinstra and Koeing equation [8]. For instance, Yasin et al. used Raman spectroscopy and confirmed the reduction of the average size of sp^2 domains and edge defects as the result of some C=C bonds reduction in graphene oxide [25].

The different Raman spectra obtained during the different reduced graphene oxide preparation steps can be ascribed to the increase of aromatic domains with a smaller size in reduced GO compared with a complete lattice structure [11, 19]. In the spectrum of reduced GO–on polyurethane (PU), identifying the Raman features of the reduced GO has proved to be difficult due to the increase in the background intensity in the region of 1300–1550 cm⁻¹. In this respect, background subtraction and quenching methods can be applied. Moving the sample position with respect to the laser beam in Raman spectroscopy allows for scanning different region of the sample (with limited lateral resolution) and can provide information about the local heterogeneity in the sample surface. For instance, spectra obtained from different areas in both GO–PU and reduced GO–PU coatings by Romani et al. were almost identical confirming the formation of homogeneous coatings in all cases [11].

3.4 X-ray Diffraction Examination

X-ray diffraction is another commonly used technique in the characterization of crystalline as well as semi-crystalline coatings. Bragg's law can be used to explain the pattern of X-rays scattered by crystals as:

$$n\lambda = 2d\sin\theta \tag{1}$$

where n, λ, d, and θ are a positive integer number, the wavelength of the incident wave, the distance between crystalline planes, and the angle of incidence for the beam, respectively.

The short wavelength that is used in X-ray diffraction, which is of the same order of the distance between the atomic or molecular structures of interest, allows studying matters at all states. Normally, X-ray diffraction (XRD) patterns are utilized in determining the distance between layers of atoms in a crystalline matter or to find the crystal structure of an unknown substance. For the XRD pattern of graphene oxide (GO) that is shown in Fig. 15, Cu Kα radiation (λ = 0.154 nm) source at a scanning rate of 1°/min has been used to obtain diffraction patterns. The characteristics peak at 26.5° corresponding to the (002) plane can be correlated to the existence of graphite. In the case of GO, the intensity of this peak immensely decreases, and a peak emerges at 10.8° which is related to the (001) plane (d = 0.78 nm). The existence of this peak in the XRD pattern of the coating suggests the successful oxidation of graphite to graphene oxide in the presence of oxidizing agents [27].

In the XRD patterns of most GO powders, a characteristic diffraction peak at about 2θ = 10.4° can be identified, which corresponds to the interlayer spacing of about 0.85 nm. The increase of interlayer spacing in a GO containing structure compared to that of graphite (0.34 nm) indicates the graphite exfoliation and the presence of oxygen-containing functional groups on GO sheets. Thus, the intensity of diffraction for this peak can be used as a characteristic for the formation of a stacked-layer structure with smaller crystallite size and a more disordered structure of the GO sheets. In general, introducing more functional groups and corrugated structure on the GO sheets with a higher degree of oxidation results in a lower intensity of XRD pattern [8].

For instance, Fig. 16 shows the XRD patterns of GO and FGO, in which GO exhibits a sharp and intense peak at 2θ = 10.92°, corresponding to a well-arranged layer structure with a d-spacing of 8.09 Å. After functionalization, the peak position for FGO changes to 2θ = 24°, corresponding to a d-spacing of 3.66 Å and the peak intensity decreases compared with GO. This reduction of the peak intensity and shift

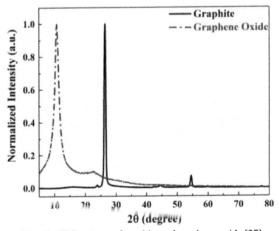

Fig. 15. XRD pattern of graphite and graphene oxide [27].

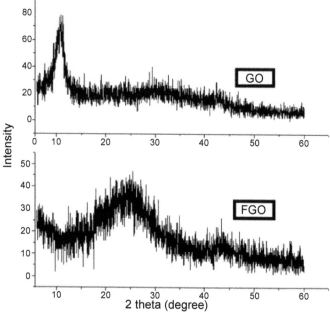

Fig. 16. XRD pattern of GO and FGO [30].

in its position confirms a decrease in oxygen-containing groups after reaction with 4-nitro aniline (NA) [30].

4. Thermal Stability Examination

Thermal stability measurement for the graphene and graphite containing coatings is another versatile method for the characterization of the coatings. Thermal stability can be determined by thermogravimetric analysis (TGA) and differential thermal analysis (DTG). The term "thermal stability" describes the thermal resistance of the compound by heating it. Representative TG-DTA curves of graphene oxide and functionalized graphene oxide samples are shown in Fig. 17a. In this figure, the decrease in weight percentage up to 85°C indicates the loss of moisture in the GO. The substantial (about 25%) weight decrease between 205°C–230°C demonstrates the loss of oxygen-containing functional groups that were introduced during oxidative treatment. In Fig. 17b, the weight loss curve of functionalized GO shows a slow weight loss (from 86 to 60 wt.%) at temperature range of 220°C to 600°C. This weight loss corresponds to the thermal decomposition of covalently bonded nitrogen-containing species. At higher temperatures (> 600°C), the weight loss is much less compared to that at lower temperatures due to the decomposition of carbonyls on the surface. From 600°C to 950°C, the observed decrease in weight loss from 60 to 50 wt.% shows that the prepared functionalized GO has an excellent thermal stability compared to that of GO [30].

TGA curves for graphene oxide and PAMAM dendrimer-modified GO (GO-PAMAM) were reported by Nonahal et al. [31] Pristine GO were more unstable and

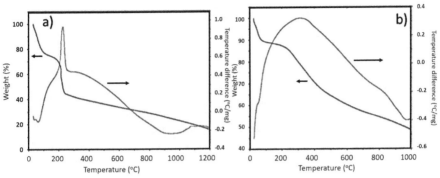

Fig. 17. TG-DTA results of (a) GO and (b) FGO [30].

displayed a significant mass loss of about 30 percent at the temperature range of 120–280°C, which was associated to the decomposition of less stable oxygen-containing functionalities. Grafting of PAMAM on GO increased the thermal stability of the modified GO with a significant weight loss (~ 30%) across the temperature range of 200–500°C. Nevertheless, below 150°C, PAMAM-grafted GO shows negligible weight loss, indicating an enhanced hydrophobicity that minimized the amount of absorbed water compared to pristine GO. The change of the hydrophobicity of the coatings directly affects its corrosion resistance properties. The adsorbed moisture usually exhibits a mass loss of about 10% below 150°C. Nonahal et al. [31] obtained the DSC thermograms of blank GO containing epoxy coatings and those containing PAMAM. In their TGA studies, different heating rates (from 5 to 20°C/min) was used. They reported a single exothermic peak after curing the unfilled epoxy system as well as for its nano-composites, signifying that the main curing reactions are governing the system over other side reactions. In all cases, by increasing the heating rate from 5 to 20°C/min, the maximum heat flow shifts towards higher temperatures. This is because a slower heating rate offers a longer time for the possible chemical reaction of the active sites in the curing system and provides sufficient time for the materials to reach equilibrium temperatures [23, 31].

TGA was also employed to study the thermal stability of waterborne polyurethane/graphene oxides nano-composites (WPU/GOs) which were made using BC4A and SC4A modified GO nanosheets. The amount of C4As bounded onto the surface of modified GOs was also measured via TGA. TGA thermograms of GO, CGO, and SGO nanosheets under the nitrogen atmosphere are shown in Fig. 18. Evaporation of adsorbed water at temperatures below 150°C lead to weight loss in all GOs, at different extents. In the case of CGO, only 9.7 wt.% loss was observed, whereas, GO and SGO exhibited 24.1 and 14.9 wt.% losses, respectively. Higher weight losses for GO and SGO below 150°C was correlated to the more hydrophilic characteristic of GO and SGO compared to CGO. From the TGA curve of GO, the principal weight loss of 62.2% can be detected at temperature range of 150°C to 300°C. This substantial weight loss corresponds to the decomposition of the less thermally stable oxygen-containing groups, including epoxide, hydroxyl, and carboxylic groups. For CGO and SGO nanosheets, in which some part of thermally instable groups is replaced by

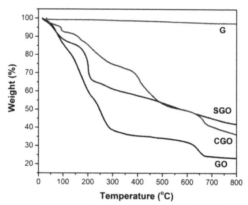

Fig. 18. TGA thermograms of prepared GO, CGO, and SGO nanosheets [6].

more thermally stable Calix arenes macrocycles [4], the weight loss at temperatures between 150°C and 300°C was lower compared to that in GO. Nevertheless, thermal stability of CGO and SGO below 300°C was higher than the stability of GO at these temperatures. The authors associated the subsequent weight loss between 300–800°C to the decomposition of the carbon skeleton and more stable phenolic groups. Based on the remaining sample weights at 800°C, the authors concluded that the CGO and SGO nanosheets contain 13.1 and 19.2 wt.% of BC4A and SC4A macrocycles, respectively [6].

Apart from the microscopic and spectroscopic characterizations of the graphene composite coatings, assessment of their performance as corrosion barrier layers should be performed in experiments directly addressing the electrochemical and chemical reactions of the coating/metal system and their surrounding environment. The following section provides examples of different fabrication methods for the formation of graphene and graphene oxide containing composite coatings, where different corrosion assessment techniques are used to characterize the corrosion resistance of the coatings under different exposure conditions.

5. Methods of Application of the Coatings and their Corrosion Examination

There are many techniques to investigate the corrosion rate of anti-corrosion coatings, including the traditional methods of calculating corrosion, polarization techniques described, and new methods of calculating corrosion [2]. Polarization techniques consist of several methods, including Tafel extrapolation, potentiodynamic measurements, cyclic polarization, and linear polarization resistance [2]. Electrochemical impedance spectroscopy (EIS) is another commonly used electrochemical method for assessment of the anti-corrosion coatings in corrosion studies. Electrochemical impedance data, in addition to information about the kinetics, also offer the possibility for postulating the corrosion mechanism. Scientists make use of EIS data to evaluate the coatings and the way they preserve metals. Electrical circuits made by components such as resistors and capacitors are utilized via the modeling procedure to explain the electrochemical properties of the coating and metal surface. Accordingly, any change in the values of the components of the electrical circuit represents the performance of a coating system [2].

In the following sections, a wide diversity of coating methods is summarized, including their feasible merits and demerits, as well as their availability for different kinds of graphene-based coatings. Eventually, depending on the coating methods of graphene-based coating, the corrosion resistance of samples are evaluated [2].

5.1 Corrosion Performance of Graphene Applied by Dip Coating

Dip coating, as a convenient and facile approach, has been extensively utilized for research purposes in the preparation of composite coatings. For instance, the corrosion protection properties of salicylaldehyde@ZIF-8 graphene oxide (szg/pvb) coating on AA2024 aluminum alloy was reported recently [34]. In this research, the synthetic SZG nano-composites at first were wet-transferred to 100 mL ethanol followed by stirring and ultrasonication. Next, 10 g PVB powder was added to the dispersion. An evenly dispersed SZG/PVB precursor solution could be obtained after 1 h rigorous stirring. Then, the pretreated aluminum alloy samples were submerged into SZG/PVB solution for 5 min, then lifted out at the rate of 100 mm/min and dried for 24 h at 30°C. These surface of these coatings were examined after immersion in 3.5% NaCl for 144 h. By examining the surface of the coated sample, the authors observed uniform corrosion and some pits on the surface. According to the results, it seems that the SZG nano-composites improve PVB coating corrosion protection properties. In the early stages of immersion (i.e., 24–96 h), SZG/PVB coating demonstrates a higher impendence value and slower degradation and only a few corrosion products were identified in the scratch of SZG/PVB coating. The better corrosion protection property on the scratched sample was associated to the active protection of SZG/PVB coating [34].

5.2 Corrosion Performance of Graphene Applied by Spray Coating

Spray coating is another coating method that is extensively used in industrial coatings, graphic art, and painting, and is the method of choice for applying thin films onto a substrate. In spray coating, many different materials with different morphologies can be used. Besides, spray coating can be used in roll-to-roll production or for fabrication of large structures [2]. For instance, novel aniline trimer (AT) functionalized graphene sheets (SAT-G) were prepared by Ye et al. [35]. The authors marked composite coatings with 0.5 and 1 wt.% of SAT-G as 0.5% SAT-G/EP and 1% SAT-G/EP, respectively. For the fabrication of composite coatings, they first dissolved powder of SAT-G in 10 mL ethyl alcohol followed by sonication for 15 min. Next, 3 g epoxy resin was added to the solution above, and the mixture was stirred for 0.5 h. After addition of 4.5 g curing agent and another 30 min. stirring, the coating was sprayed on the preprocessed mild steel. The authors recorded the EIS data of as-prepared samples during 75 days immersion in 3.5 wt% NaCl. The observed large diameter of the capacitive arc in their obtained EIS results suggested an excellent barrier property of the fabricated coatings [35]. The hybrid functionalized graphene sheets that were prepared in this study were well dispersed in the epoxy matrix, providing a superior anti corrosion properties to it with promising self healing and shielding effects.

In another study, Pourhashem et al. investigated the effect of silane functionalized graphene quantum dots (f-GQDs) on the corrosion resistance of solvent-based epoxy coatings [36]. The preparation of nano-composite epoxy/f-GQDs coatings comprised dispersion of f-GQDs in acetone using ultrasonication bath. After mixing the dispersion with a polyamide hardener under sonication, the mixture was stirred and heated to 50°C to allow the solvent evaporation. Next, the prepared mixture was mixed with a stoichiometric amount of epoxy resin. Eventually, the prepared nanocomposites were sprayed on sandblasted mild steel substrates. The authors observed that the sample containing epoxy/f-GQDs composite coatings is more resistant to corrosion than the epoxy-only coatings. The epoxy coating containing 0.5 wt.% f-GQDs exhibited higher protection capability and lower corrosion rate than epoxy coating containing only 0.1 wt.% f-GQDs, which confirms the positive role of f-GQDs on the corrosion resistance that is provided in the composite coatings. Based on the results of open circuit potential (OCP) values of samples during 28 days of immersion in 3.5 wt.% NaCl solution, the authors observed that nano-composite epoxy/f-GQDs coatings show more positive OCP values than pure epoxy coatings. With prolonged immersion in 3.5 wt.% NaCl solution, the OCP of samples moves to more negative values, which was attributed to the diffusion of corrosive agents through coating to coating/metal interface [36, 37].

In another research, graphene oxide sheets with different aspect ratios were first prepared by slightly controlling the procedure of chemical peeling which then were used to fabricate GO/epoxy composite coatings [8]. The corresponding potentiodynamic polarization (PDP) curves of steel substrate coated by pure epoxy coating as well as those containing GO of different aspect ratios (GO-a/EP, GO b/EP, and GO-c/EP) are shown in Fig. 19.

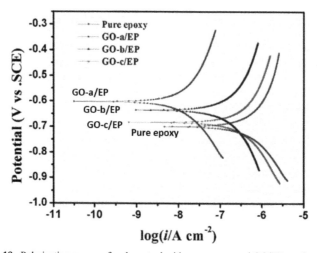

Fig. 19. Polarization curves of carbon steel with pure epoxy and GO/EP coatings [8].

As can be observed in this figure, compared to pure epoxy coating, corrosion potential for GO/EP composite coatings changes to more positive values, and the observed shifts in i_{corr} toward lower values indicate high corrosion resistance for epoxy composite coatings. Moreover, the GO-a/EP coating exhibit the most positive corrosion potential (–0.603 V) and the lowest corrosion current density (1.164 × 10^{-8} Amp/cm^2) among the studied samples. The better corrosion performance of GO-a sheets was correlated to the high aspect ratio graphene oxides in its structure. The authors suggested that composite coatings containing GO-a provide a substantial barrier against corrosion by blocking the diffusion path for corrosive ions toward the substrate. According to the PDP results in Fig. 19, the calculated corrosion rates were 5.90 × 10^{-3} mm/year (for pure epoxy), 1.35 × 10^{-4} mm/year (for GO-a/EP), 1.64 × 10^{-3} mm/year (for GO-b/EP), and 3.62 × 10^{-3} mm/year (for GO-c/EP) [8].

In another study by Selim et al., a novel and robust alkyd/exfoliated graphene oxide nanosheet composite coating was deposited on substrate via solution casting and spray coating [16]. The process of preparing this coating was such that different GO nanofiller concentrations (0.05, 0.1, 0.5, 0.1, 2.5, and 5 wt.%) were dispersed in oil. The mixture of SFO-alkyd (87 wt.%-nanofiller) GO nanosheets (0.05–5 wt.%) in fat (10 wt.%) was then ultrasonicated for 15 min to form different nanocomposite solutions. The unfilled and SFO-alkyd/GO sheets composites were sprayed on the substrate of steels. Salt-spray tests were used to assess the corrosion protection of the coatings, as can be seen in Fig. 20. According to these results, well-distributed GO nanosheets (0.5 wt.%) filled alkyd formulation exhibited more resistance to corrosion than unfilled alkyd formulation. By contrast, higher nanosheet loadings up to 5 wt.% showed increased corrosion as a result of nanofiller aggregation, reduced cohesion between the substrate and nano-composite coating as well as increased penetrability of corrosive media. The observed data confirmed the significance of the proper-distribution of GO nanosheets in the SFO-alkyd for coating corrosion protection applications [16, 38].

In another work regarding spray coating of the graphene-based nanocomposite, the GO-Fe$_3$O$_4$@ poly (DA+KH550) hybrid/epoxy composite coating was prepared in the following way: a certain amount of curing agent and GO-Fe$_3$O$_4$@poly (DA+KH550) hybrid with various loading contents were dispersed into the epoxy resin and the composites were evenly sprayed on the steel surface [20]. The possible anti-corrosion mechanism offered by such composite coating is illustrated in Fig. 21 [24]. According to this figure, it can be deduced that in the presence of epoxy coating, O$_2$ and H$_2$O can potentially penetrate through the micropores of the coating and reduce its functionality. Yet, with the presence of graphene within the epoxy matrix, coating performance is improved. With the addition of 0.5 wt.% GO-Fe$_3$O$_4$ hybrid (Fig. 21), the anti-corrosion behavior of the coating was greatly enhanced due to the high aspect ratio of graphene oxide and the synergistic effect of graphene oxide and Fe$_3$O$_4$ nanoparticles. Comparable performance in corrosion behavior was also obtained by combining graphene oxide/Al$_2$O$_3$, graphene oxide/silica, and graphene/TiO$_2$ hybrids. Zhang et al. developed a coating based on a novel bio-inspired co-modification of graphene oxide/Fe$_3$O$_4$ hybrids by dopamine and silane KH550. The surface modificatoion of oxide/

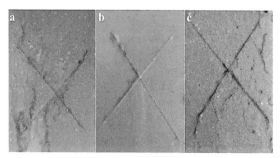

Fig. 20. Salt spray results on the coatings after 500 h of exposure; (a) unfilled SFO-alkyd; (b) excellent distributed SFO-alkyd/GO nanosheets (0.5 wt.%) and (c) agglomerated SFO-alkyd/GO nanosheets (5 wt.%) [16].

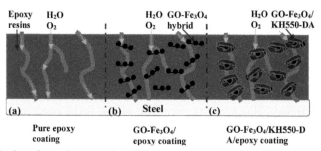

Fig. 21. Mechanism of corrosion protection of (a) pure epoxy coating, (b) GO-Fe$_3$O$_4$/epoxy coating, and (c) GO-Fe$_3$O$_4$@poly (KH550+DA) hybrid/epoxy coating [24].

Fe$_3$O$_4$ hybrid occured by the self-polymerization between dopamine and secondary functionalized monomer (KH550). Many functional groups such as -NH$_2$ and -OH groups were obtained after co-modification, which would improve their dispersion in the epoxy matrix. Moreover, these functional groups participated in the curing reaction of the epoxy during the sample preparation, a mechanism which further enhances the interfacial adhesion between the nanoparticles and epoxy matrix, as shown in Fig. 21c [24].

In another work using the above-mentioned method, it was found that nano-composite coatings exhibit lower OCP values compared to pure epoxy samples. It was also argued that the OCP values of epoxy/0.05 PG (polycrystalline graphene) sample are permanently more positive than the other coatings during immersion in NaCl containing solution [18]. The authors found an increasing order of the OCP values as: epoxy/0.05 PG > epoxy/0.1 PG > epoxy/0.3 PG > epoxy/0.5 PG > pure epoxy. The reason for the increased OCP is the better performance of coatings with higher PG content against the penetration of aggressive ions and their stronger adhesion to the substrate. Also, the authors observed a reduction of OCP levels of all specimens with increasing immersion time [18]. Overall, the electrochemical methods proved to be facile and powerful tools to assess the corrosion protection efficiency of the graphene and graphite containing coatings that were fabricated via deep coating or spray coating.

5.3 Corrosion Performance of Graphene Applied by Electrochemical Deposition

Another method of applying coatings to substrates is the electrochemical deposition method, which is much easier to use than other methods. This method can be carried out at a low ambient temperature and high deposition rates. The low cost of the electrodeposition method compared to other methods [2, 25] is considered as another advantage for preparation of the graphene and graphite containing coatings to improve the corrosion resistance. For instance, bright nickel–graphene nanocomposite coatings have been successfully fabricated onto the mild steel by electrochemical co-deposition technique. The sample deposited at a current density of 9 Amp/dm^2 showed higher corrosion resistance in 3.5 wt.% NaCl solution compared to bright nickel coatings without graphene. The results of the EIS spectrum show a higher impedance at a low density of 9 Amp/dm^2. It was argued that the corrosion behavior of nickel–graphene composite coatings highly depends on the distribution of graphene sheets within the nickel-metal matrix and that the homogeneous graphene distribution in the nickel matrix during the electrostatic process can lead to an increase in corrosion resistance of the coatings. Essentially, well distributed nanosheets in the nickel coatings can fill the holes and cavities in the nickel matrix, and consequently prevent the development of defects in the composite coatings [25, 28].

Xiaong et al. utilized electroless Cu coating bath composed of 10 gr/L CuSO$_4$·5H$_2$O, 15 gr/L Na$_3$C$_6$H$_5$O$_7$·2H$_2$O, 28 gr/L NaH$_2$PO$_2$·H$_2$O, 30 gr/L H$_3$BO$_3$, and 1 gr/L NiSO$_4$·7H$_2$O. The pH value was adjusted in the range of 9–9.5, and the temperature was maintained at 65°C in their study. The GO aqueous solution was ultrasonically dispersed for 1 hr before plating for GO homogenization in the electroless bath. During plating, the electroless bath was continuously stirred by a magnetic stirrer to avoid aggregation. The duration of the plating was 1 h. Corrosion potential (E_{corr}) of the tested specimens for substrate was –0.889 V and for pure Cu coating was –0.592 V. This value for Cu-Gr composite coatings with (20–60–100) mg/L GO concentrations was –0.571 V, –0.571 V, –0.447 V, respectively. Minimum i_{corr} value was also found at the 60 mg/L GO content in the bath. It was found that the Cu-Gr coating could significantly improve corrosion resistance in Cl$^-$ solution, and the best effect was obtained at the GO adding amount around 60 mg/L in the electroless bath [34]. The authors found that both conventional Cu coatings and Cu-Gr composite coating effectively inhibited the penetration of Cl$^-$ toward the iron substrate. Many researchers have recommended that when Cu is exposed to neutral and aerated Cl$^-$ media, the passivation layer is comprised mainly of Cu$_2$O. For the Cu-Gr coatings, the authors proposed two mechanisms for enhancing the corrosion resistance: the refined grains and dense structure of the coating as a result of the Gr addition, and the high impermeability of Gr which serves as corrosion protection barriers to impede the Cu$^+$ ion diffusion across the coating [34, 39].

5.4 Corrosion Performance of Graphene Applied by CVD and LBL

Chemical vapor deposition (CVD) is another method of covering graphene-based nanocomposite coatings and has been used by many researchers in recent articles.

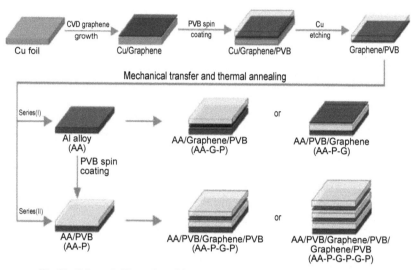

Fig. 22. Schematic illustration of the preparation of nanocomposite coating [40].

CVD is a desirable method for production of graphene in large scales, which provides the conditions for producing large-scale graphene coatings as well as pure and dense coatings [2]. For instance, when AA2024-T3 is coated with a polymer/SLGr/polymer/SLGr/polymer coating, a relatively long-term (120 days) corrosion protection can be achieved. Measurements for the bare samples, AA-P-P-P, and AA-P-G-P-G-P (See Fig. 22) after 1, 30, or 120 days of immersion in 3.5 wt.% NaCl solution was reported. When P-P-P coatings cover AA, the corrosion rate of the coated sample, after one day immersion, is two orders of magnitude lower than that of the bare AA (20 nm/year for AA-P-P-P vs. 4 µm/year for naked AA). Nevertheless, the corrosion rate of P-P-P coated sample increases with immersion time (0.3 µm/year for AA-P-P-P at 30 days vs. 20 nm/year for AA-P-P-P at one day), which is an indication of the degradation of P-P-P coating. Furthermore, the potentiodynamic response for AA-P-G-P-G-P exhibit low current values and corrosion rates are below 2 nm/year. Further immersion of the nanocomposite coatings (for 120 days) also results in degradation of the coatings. Overall, a good agreement can be obtained on these systems between the polarization measurements and the EIS results [40].

5.5 *Direct Application and Curing*

This process is a suitable method in which a coating can be applied on the substrate under ambient conditions, e.g., curing occurs at room temperature. Therefore, this approach is appropriate solely for the coatings that can be cured under ambient conditions, such as UV-curable nano-composite coatings [2]. For instance, graphene oxide–poly(urea-formaldehyde) composites (GUF) which were used for corrosion protection of mild steel were fabricated by a simple processes described by Zhang et al. [41]. EIS measurements were used to evaluate and compare the effects of the five different GUF composites on the barrier performance and corrosion protection of epoxy coatings. The GUF/EP coatings that Zheng et al. investigated (see Table 2 for more details) showed higher impedance values than pure epoxy or GO/EP counterparts. Among these coatings, only the one with middle concentration of GO/EP exhibited the absolute impedance close to 90° at a frequency of 0.01 Hz. After 170 h immersion, the $|Z|$ 0.01 Hz of the studied epoxy and GO/EP coatings decreased to 1.1×10^9 and 9.5×10^9 $\Omega \cdot cm^2$, respectively. The $|Z|$ 0.01 Hz values of composite coatings with the lowest and highest GO fell to 4.5×10^{10} and 6.1×10^{10} $\Omega \cdot cm^2$, respectively. The changes of $|Z|$ for the prepared coatings after immersion in the corrosive medium for 2000 h also was different for the prepared composite coatings, depending upon the concentration of GO in their structure. Overall, the GUF composites exhibited better corrosion protection than only GO sheets in the epoxy coating, findings that were consistent with those in previous research. The optimal corrosion protection performance of GUF composites was obtained with 8.6 wt.% GO sheets [41] in the composite coating.

WPU/GOs composite coatings with BC4A and SC4A also showed excellent anti-corrosion properties, based on the electrochemical evaluations [6]. The procedure of application of these coatings is schematically shown in Fig. 23.

The authors quantified the corrosion performance of uncoated and coated mild steels after one-week exposure to 3.5 wt.% NaCl solution, by potentiodynamic polarization (PDS) technique. The percentage of inhibition efficiency (*IE*) as a useful parameter to evaluate the anti-corrosion efficiency of WPU and WPU/GOs nano-composite coatings was also calculated using the formula given below [6]:

$$percentage \ of \ IE = \frac{I_{corr \ (0)} - I_{corr \ (1)}}{I_{corr \ (0)}} \times 100 \qquad (2)$$

where $i_{corr \ (0)}$ and $i_{corr(i)}$ are the corrosion current values for uncoated and coated mild steels, respectively. Based on the results of this study, I_{corr} values for the coated mild steels were much lower than that for uncoated steel. Furthermore, the E_{corr} of the WPU and WPU/GOs coated mild steels increases toward more positive potentials compared to that of uncoated steel. Analyzing the results of this study indicates an improvement in corrosion resistance of WPU and WPU/GOs coated

Table 2. Compositions of the coatings as prepared [41].

Coating	Epoxy	GO/EP	GUF1/EP	GUF2/EP	GUF3/EP	GUF4/EP	GUF5/EP
Epoxy resin/wt.%	66.7	66.5	65.6	65.6	65.6	65.6	65.6
Hardener/wt.%	33.3	33.2	32.8	32.8	32.8	32.8	32.8
Filler/wt.%	0 0.3	1.6	1.6	1.6	1.6	1.6	1.6

WPU   GO and C4As-GOs   WPU/GOs   Coated mild steels

Fig. 23. Schematic illustration of preparation steps for the coating [6].

mild steels in comparison with their uncoated counterpart. This improved corrosion resistance property was correlated with the barrier effect of coatings on the surface of mild steels. Among the WPU/GOs samples, WPU/CGO and WPU/SGO samples containing SC4A as modified GOs (CGO and SGO nanosheets) provided the best anti-corrosion properties on mild steel [6].

Ramezanzadeh et al. applied defect-free epoxy coatings without and with neat GO and GO-polyanalyne (PAni)-CeO$_2$ particles on mild steel. Their results indicated that the corrosive electrolyte diffusion into the coating matrix through porosities and cavities results in the deterioration of the coating barrier performance. However, in their EIS measurements, they found a time independent capacitive behavior in the epoxy coating filled with GO-PAni-CeO$_2$ particles, over a wide range of frequencies for the epoxy coating filled with GO-PAni-CeO$_2$ particles, the coating film showed capacitive behavior in a wide range of frequency during the EIS measurements and did not change during the immersion. According to the salt spray test results given in Fig. 24, as the immersion time increased, the coatings degraded, lost their adhesion to the substrate, and corrosion products appeared on their surface [22].

Characterization and corrosion protection of reduced GO coating reinforced with waterborne soy alkyd nano-composites was performed by Irfan et al. [42]. They prepared 80 wt.% solutions of WSA-PMF-80 and WSA-PMF-reduced GO-x in ethanol in a reaction vessel using Poly(melamine-co-formaldehyde) isobutylated solution (PMF) cross-linker as a curing agent. These reaction mixtures were ultrasonicated for 1 h at 50°C after contentious stirring for 2.5 h at 120–130°C. Finally, these solutions were applied by brush techniques on the prepared surface of

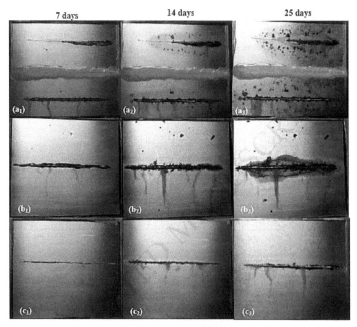

Fig. 24. Salt spray test result for the (a1, a2, a3) neat epoxy, (b1, b2, b3) GO/Epoxy, and (c1, c2, c3) GO-PANI-CeO$_2$/Epoxy samples after 25 days [22].

CS (carbon steel) specimens. The authors suggested, based on the increase in E_{corr} and the decrease in i_{corr} values, that in the case of composite coatings the nano-composite coating provides excellent protection to CS via the physical barrier mechanism. The higher value of E_{corr} based on their EIS results indicated superior barrier property of the fabricated coatings [42]. The authors observed a decreasing trend in the corrosion protection efficiency of the fabricated coatings as: CS < WSA-PMF-80 < WSA-PMF-80-RGO-0.5 < WSA-PMF-80-RGO-1.0 < WSA-PMF-80-RGO-1.5 [42].

5.6 Corrosion Performance of Graphene Applied by Spin Coating

Spin coating is another widespread and commonly utilized method for the rapid deposition of thin-film layers onto planar substrates. In a spin coater device usually, rotating stage and vacuum pumps are utilized to hold the substrate to enable covering the surface of the sample with a uniform and thin film by drop-casting a dispersion coating solution onto the surface [2]. Nayak et al. studied the corrosion protection performance of functionalized graphene oxide nanocomposite coating prepared via spin coating method on mild steel [30]. FGO composite coating showed higher (more positive) OCP value compared to that of EP and blank samples. The authors correlated the barrier property of the prepared FGO composite in epoxy coating to the decrease of the diffusion of electrolyte toward the steel surface [30]. From their impedance of the samples, it could be inferred that the nano-composite of graphene has a higher impedance value compared to that of EP and blank. After some days of immersion in 3.5 wt.% NaCl solution, the $Z_{0.01Hz}$ value of (FGO+EP) decreases significantly [30]. PTCA-G was prepared according to the previously reported work and the Nyquist plots of bare substrate and coating without additives and with different additives were reported by Yang et al. [15]. The authors found that the diameter value of semicircles in the Nyquist spectra followed the sequence of bare substrate < pure EP < PTCA/EP < G/EP < PTCA-G/EP and the EP containing PTCA-G composite showed the best anti-corrosion performance. The corresponding Nyquist an Bode plots of coatings with different volume ratios of PTCA-G composite are provided in Figure 25 [15].

The corrosion mechanism of reduced GO coating on a magnesium substrate was studied by Chu et al. [20]. Magnesium has passive layers mainly composed of (MgO and Mg (OH)$_2$), which lose their protective properties against aggressive ions such as chloride ions. The pure reduced GO coating cannot form a continuous phase because of its weak interaction between the neighboring GO units and thus a large number of capillaries can form between graphene sheets. Essentially, the oxidized regions of original GO laminates have large spacing (d), which leaves a space between the non-oxidized regions, which allow the formation of capillary network in a repeated self assembly process. This network provides paths for penetration of Cl- and water. Furthermore, the porosity in the structure of reduced GO coating facilitates the electrolyte access to the substrate, which accelerates galvanic corrosion between graphene sheets and Mg alloy. Finally, the reduced GO coating gradually loses its barrier and protection effect, as shown in Fig. 26a. The role of polyvinyl alcohol (PVA) in these coatings is very critical because PVA increases the adhesion of the coating as a result of hydrogen bonds between reduced GO sheets and PVA chains. This cross-linking results in the formation of a uniform film. Furthermore, the

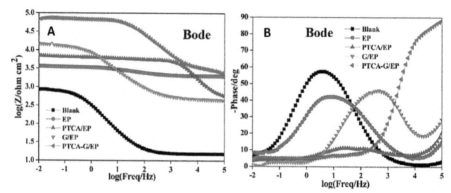

Fig. 25. Bode (A) magnitude (B) phased of coatings with different volume ratios of PTCA-G in 3.5 wt.% NaCl solution [15].

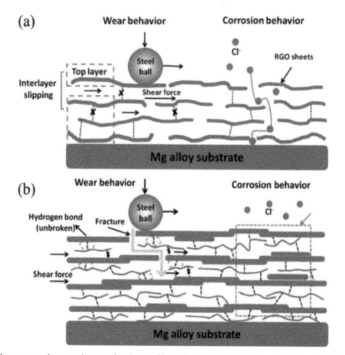

Fig. 26. The proposed protection mechanisms offered by hybrid coatings on the Mg substrate during the corrosion process. (a) Pure RGO coating. (b) RGO/PVA hybrid coating [20].

branches of PVA molecules can order into the nonoxidized regions of reduced GO sheets, which efficiently decrease the intrinsic defects and facilitate the formation of an integrated and continuous film during deposition, which in turn improves the corrosion resistance of coating/Mg substrate. The degree of ordering and compactness of reduced GO/PVA hybrid coating can be affected by the content of the PVA. The corresponding corrosion mechanism has been summarized in Fig. 26b [20, 43].

5.7 Corrosion Performance of Graphene Nanocomposite Applied by Other Methods

Apart from the previously mentioned methods for fabrication of graphene or graphite containing composite coating, other methods such as drop-casting, cold-pressing, and wire-wound lab rod have also been utilized by different groups. A methodology for producing nanocomposite films of polyurethane and graphene oxide and polyurethane and reduced graphene oxide was reported [11]. Figure 27 presents the Nyquist plots obtained for pure PU, GO–PU, and reduced GO–PU composite coatings. This figure demonstrates the enhancement of the barrier properties after the incorporation of reduced GO into the PU matrix, which partially prevents the penetration of the corrosive species through the coating. In these composite coatings, localized corrosion was revealed by immersion of coated steel in corrosive media. The authors correlated the increase in the $(R_{cor} + R_{po})$ resistances values from 170 MΩ cm^2 for PU coating to 262 MΩ cm^2 and 504 MΩ cm^2 for GO–PU and reduced GO–PU, respectively. It was hence concluded that reduced GO–PU coatings are more useful for protecting against corrosion [11, 12]. Based on all observations, the performance of the reduced GO–PU coatings were comparable to (or even better than) those obtained for functionalized graphene or functionalized graphene oxide nanocomposite coatings. It could also be concluded that the longer the penetration of the aggressive ions is delayed, the better will be the corrosion resistance of the graphene-based nanocomposite coating [11, 12].

The graphene encapsulated Cu composite was subjected to EIS evaluations in another study [32]. The corrosion performance of the coated specimens was assessed based on the Bode modulus plot for Gr/Cu-ip and Gr/Cu-cp, wherein the former corrosive ions penetrated more graphene layers in the so-called in-plane direction and in the latter (called cross-plane) more copper channels were available. A different corrosion mechanism for each coating was deduced based on the maximum time constant showing a sequence of Gr/Cu-ip < Gr/Cu-cp < pure copper, confirming the better corrosion resistance of Gr/Cu-ip. Beside the corrosion resistance properties of the graphene containing composite coatings, other coating properties can also be modified during the coating deposition. For instance, Tsai et al. showed that by

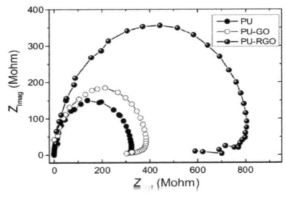

Fig. 27. Nyquist plots for PU, GO–PU, and reduced GO–PU coatings [11].

applying the wire-wound method for fabrication of graphene containing coatings, the adhesion property of the graphene/PU composite coatings was enhanced by adding 4 wt.% and 8 wt.% of graphene [12].

In another study, the reduced GO nanocomposite with different contents of oxygen-containing functional groups were applied using a film applicator [19]. For this purpose, the GO nanosheets were filled into the epoxy matrix for corrosion protection. The authors showed that the impedance of the neat epoxy coating was strikingly low, which allowed following a time constant in the low-frequency range after 30 days immersion in 3.5 wt.% NaCl (Fig. 28c–d). When immersion

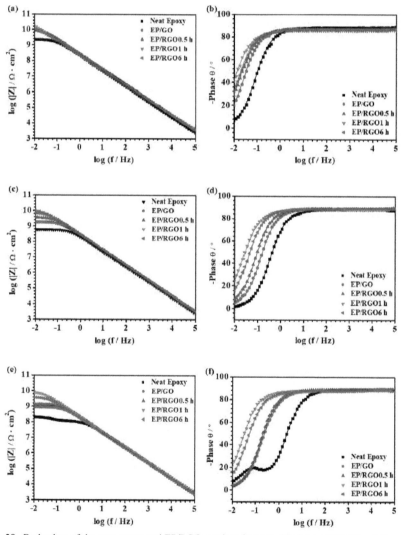

Fig. 28. Bode plots of the neat epoxy and EP/RGO coatings immersed in 3.5 wt.% NaCl solution after (a, b) 14 days, (c, d) 30 days, and (e, f) 60 days; solid lines and marker points represents the fitted and experimental data, respectively [19].

was extended to 60 days, two-time constants appear in the Bode phase diagram of neat epoxy (Fig. 28e–f). Overall, EP/reduced GO coatings exhibit proper corrosion protection performance, owing to their hydrophobic property and outstanding impenetrability associated with the low amount of oxygen-containing functional groups in their structure [19].

6. Concluding Remarks

In this chapter, we summarized the various methods for application and characterization of graphene and graphene nanocomposite coatings. We covered the recent literature that considered graphene-based nanocomposite coatings for improving the corrosion resistance of the substrate and described the utilized method for application of these coatings on various metallic substrates. The corrosion resistance properties of these composited coatings were discussed based on the individual analytical assessments and the main observations were discussed. It is clear that the complete assessment of the complex coatings requires the simultaneous application of two or more of these analytics. In the assessment of the structure of the graphene-based nanocomposites, microscopic evaluations using SEM were found to be almost always necessary. SEM investigations allow for the detection of the shape of phases, microstructure, morphology, dispersion, or agglomeration of the second phase in the matrix and enable post-analysis of the fracture morphology in the case of sample damage. Within spectroscopic tools, EDS analysis was the most promising, broadly used technique which enables the detection of the proportion of elements in the structure of the composite coatings as well as the dispersion of coating elements within the matrix. The FTIR is also quite essential for the characterization of the functional groups in the compounds and enables following the coatings modifications before and after the chemical treatment. XRD test also was found to be very helpful for the determination of the crystalline structure of the synthesized coating.

We provided examples where the role of graphene nanosheets on the fabrication of a high-performance corrosion protection system was studied by EIS, PDP, and salt spray test. In our literature survey, we found that the spray coating was used more than any other method for the fabrication and deposition of the composite coatings.

In brief, in this chapter, we introduced a sequence of analytical characterization methods that can be utilized for in-depth characterization of the graphene-based composite coatings, *in situ* and *ex situ*. It is expected that with the advancement of the high-pressure XPS, SEM and introduction of the *in situ* TEM, as well as functional scanning probe techniques, even more fundamental and mechanistic information on the role of each coating constituents in the composite coatings in improving the corrosion resistance of the sample/coating can be obtained in the near future.

7. Acknowledgments

Alireza Basavandmanesh and Hamed Kazemi are acknowledged for general discussions and updating the references.

References

[1] Paszkiewicz, S. and A. Szymczyk. 2019. Chapter 6—Graphene-based nanomaterials and their polymer nanocomposites. pp. 177–216. *In*: Karak, N. (ed.). Nanomaterials and Polymer Nanocomposites. Elsevier.

[2] Kakaei, K., M. D. Esrafili and A. Ehsani. 2019. Chapter 8—Graphene and anticorrosive properties. pp. 303–337. *In*: Kakaei, K., M. D. Esrafili and A. Ehsani (eds.). Interface Science and Technology. Elsevier.

[3] Hosseinpour, S., J. Hedberg, S. Baldelli, C. Leygraf and M. Johnson. 2011. Initial oxidation of alkanethiol-covered copper studied by vibrational sum frequency spectroscopy. The Journal of Physical Chemistry C 115: 23871–23879.

[4] Hosseinpour, S., M. Göthelid, C. Leygraf and C. M. Johnson. 2013. Self-assembled monolayers as inhibitors for the atmospheric corrosion of copper induced by formic acid: A comparison between hexanethiol and hexaneselenol. Journal of the Electrochemical Society 161: C50.

[5] Gretić, Z. H., E. K. Mioč, V. Čadež, S. Šegota, H. O. Ćurković and S. Hosseinpour. 2016. The influence of thickness of stearic acid self-assembled film on its protective properties. Journal of the Electrochemical Society 163: C937.

[6] Mohammadi, A., M. Barikani, A. H. Doctorsafaei, A. P. Isfahani, E. Shams and B. Ghalei. 2018. Aqueous dispersion of polyurethane nanocomposites based on calix [4] arenes modified graphene oxide nanosheets: Preparation, characterization, and anti-corrosion properties. Chemical Engineering Journal 349: 466–480.

[7] Mohan, V. B., K. -t. Lau, D. Hui and D. Bhattacharyya. 2018. Graphene-based materials and their composites: A review on production, applications and product limitations. Composites Part B: Engineering 142: 200–220.

[8] Jiang, F., W. Zhao, Y. Wu, J. Dong, K. Zhou, G. Lu and J. Pu. 2019. Anti-corrosion behaviors of epoxy composite coatings enhanced via graphene oxide with different aspect ratios. Progress in Organic Coatings 127: 70–79.

[9] Jafari, Y., S. M. Ghoreishi and M. Shabani-Nooshabadi. 2016. Electrochemical deposition and characterization of polyaniline-graphene nanocomposite films and its corrosion protection properties. Journal of Polymer Research 23: 91.

[10] Lingamdinne, L. P., Y. -L. Choi, I. -S. Kim, J. -K. Yang, J. R. Koduru and Y. -Y. Chang. 2017. Preparation and characterization of porous reduced graphene oxide based inverse spinel nickel ferrite nanocomposite for adsorption removal of radionuclides. Journal of Hazardous Materials 326: 145–156.

[11] Romani, E. C., S. Nardecchia, C. Vilani, S. Qi, H. Dong and F. L. Freire. 2018. Synthesis and characterization of polyurethane/reduced graphene oxide composite deposited on steel. Journal of Coatings Technology and Research 15: 1371–1377.

[12] Tsai, P. -Y., T. -E. Chen and Y. -L. Lee. 2018. Development and characterization of anticorrosion and antifriction properties for high performance polyurethane/graphene composite coatings. Coatings 8: 250.

[13] Egerton, R. F. 2005. Physical Principles of Electron Microscopy. Springer.

[14] Haugstad, G. 2012. Atomic Force Microscopy: Understanding Basic Modes and Advanced Applications. John Wiley & Sons.

[15] Yang, T., Y. Cui, Z. Li, H. Zeng, S. Luo and W. Li. 2018. Enhancement of the corrosion resistance of epoxy coating by highly stable 3, 4, 9, 10-perylene tetracarboxylic acid functionalized graphene. J. Hazard Mater. 357: 475–482.

[16] Selim, M. S., S. A. El-Safty, M. A. Shenashen, M. A. El-Sockary, O. M. A. Elenien and A. M. El-Saeed. 2019. Robust alkyd/exfoliated graphene oxide nanocomposite as a surface coating. Progress in Organic Coatings 126: 106–118.

[17] Ghauri, F. A., M. A. Raza, M. S. Baig and S. Ibrahim. 2017. Corrosion study of the graphene oxide and reduced graphene oxide-based epoxy coatings. Materials Research Express 4: 125601.

[18] Pourhashem, S., A. Rashidi, M. R. Vaezi, Z. Yousefian and E. Ghasemy. 2018. The effect of polycrystalline graphene on corrosion protection performance of solvent based epoxy coatings: Experimental and DFT studies. Journal of Alloys and Compounds 764: 530–539.

[19] Wang, M. -H., Q. Li, X. Li, Y. Liu and L. -Z. Fan. 2018. Effect of oxygen-containing functional groups in epoxy/reduced graphene oxide composite coatings on corrosion protection and antimicrobial properties. Applied Surface Science 448: 351–361.

[20] Chu, J., L. Tong, J. Zhang, S. Kamado, Z. Jiang, H. Zhang and G. Sun. 2019. Bio-inspired graphene-based coatings on Mg alloy surfaces and their integrations of anti-corrosive/wearable performances. Carbon 141: 154–168.

[21] Hikku, G., K. Jeyasubramanian, A. Venugopal and R. Ghosh. 2017. Corrosion resistance behaviour of graphene/polyvinyl alcohol nanocomposite coating for aluminium-2219 alloy. Journal of Alloys and Compounds 716: 259–269.

[22] Ramezanzadeh, B., G. Bahlakeh and M. Ramezanzadeh. 2018. Polyaniline-cerium oxide (PAni-CeO$_2$) coated graphene oxide for enhancement of epoxy coating corrosion protection performance on mild steel. Corrosion Science 137: 111–126.

[23] Mooss, V. A., A. A. Bhopale, P. P. Deshpande and A. A. Athawale. 2017. Graphene oxide-modified polyaniline pigment for epoxy based anti-corrosion coatings. Chemical Papers 71: 1515–1528.

[24] Zhan, Y., J. Zhang, X. Wan, Z. Long, S. He and Y. He. 2018. Epoxy composites coating with Fe$_3$O$_4$ decorated graphene oxide: Modified bio-inspired surface chemistry, synergistic effect and improved anti-corrosion performance. Applied Surface Science 436: 756–767.

[25] Yasin, G., M. Arif, M. Shakeel, Y. Dun, Y. Zuo, W. Q. Khan, Y. Tang, A. Khan and M. Nadeem. 2018. Exploring the nickel–graphene nanocomposite coatings for superior corrosion resistance: manipulating the effect of deposition current density on its morphology, mechanical properties, and erosion-corrosion performance. Advanced Engineering Materials 20: 1701166.

[26] Wang, H., Y. He, G. Fei, C. Wang, Y. Shen, K. Zhu, L. Sun, N. Rang, D. Guo and G. G. Wallace. 2019. Functionalizing graphene with titanate coupling agents as reinforcement for one-component waterborne poly(urethane-acrylate) anticorrosion coatings. Chemical Engineering Journal 359: 331–343.

[27] Asgar, H., K. Deen, Z. U. Rahman, U. H. Shah, M. A. Raza and W. Haider. 2019. Functionalized graphene oxide coating on Ti6Al4V alloy for improved biocompatibility and corrosion resistance. Materials Science and Engineering: C 94: 920–928.

[28] Jabbar, A., G. Yasin, W. Q. Khan, M. Y. Anwar, R. M. Korai, M. N. Nizam and G. Muhyodin. 2017. Electrochemical deposition of nickel graphene composite coatings: effect of deposition temperature on its surface morphology and corrosion resistance. RSC Advances 7: 31100–31109.

[29] Khobragade, N., K. Sikdar, B. Kumar, S. Bera and D. Roy. 2019. Mechanical and electrical properties of copper-graphene nanocomposite fabricated by high pressure torsion. Journal of Alloys and Compounds 776: 123–132.

[30] Nayak, S. R. and K. N. S. Mohana. 2018. Corrosion protection performance of functionalized graphene oxide nanocomposite coating on mild steel. Surfaces and Interfaces 11: 63–73.

[31] Nonahal, M., H. Rastin, M. R. Saeb, M. G. Sari, M. H. Moghadam, P. Zarrintaj and B. Ramezanzadeh. 2018. Epoxy/PAMAM dendrimer-modified graphene oxide nanocomposite coatings: Nonisothermal cure kinetics study. Progress in Organic Coatings 114: 233–243.

[32] Jin, B., D. -B. Xiong, Z. Tan, G. Fan, Q. Guo, Y. Su, Z. Li and D. Zhang. 2019. Enhanced corrosion resistance in metal matrix composites assembled from graphene encapsulated copper nanoflakes. Carbon 142: 482–490.

[33] Wu, W., J. Liu, X. Li, T. Hua, X. Cong, Z. Chen, F. Ying, W. Shen, B. Lu and K. Dou. 2018. Incorporation graphene into sprayed epoxy–polyamide coating on carbon steel: corrosion resistance properties. Corrosion Engineering, Science and Technology 53: 625–632.

[34] Xiong, L., J. Liu, M. Yu and S. Li. 2019. Improving the corrosion protection properties of PVB coating by using salicylaldehyde@ZIF-8/graphene oxide two-dimensional nanocomposites. Corrosion Science 146: 70–79.

[35] Ye, Y., D. Zhang, T. Liu, Z. Liu, J. Pu, W. Liu, H. Zhao, X. Li and L. Wang. 2019. Superior corrosion resistance and self-healable epoxy coating pigmented with silanzied trianiline-intercalated graphene. Carbon 142: 164–176.

[36] Pourhashem, S., E. Ghasemy, A. Rashidi and M. R. Vaezi. 2018. Corrosion protection properties of novel epoxy nanocomposite coatings containing silane functionalized graphene quantum dots. Journal of Alloys and Compounds 731: 1112–1118.

[37] Arthisree, D. and G. M. Joshi. 2017. Study of polymer graphene quantum dot nanocomposites. Journal of Materials Science: Materials in Electronics 28: 10516–10524.

[38] Zhu, K., X. Li, H. Wang, J. Li and G. Fei. 2017. Electrochemical and anti-corrosion behaviors of water dispersible graphene/acrylic modified alkyd resin latex composites coated carbon steel. Journal of Applied Polymer Science 134.

[39] Alhumade, H., A. Abdala, A. Yu, A. Elkamel and L. Simon. 2016. Corrosion inhibition of copper in sodium chloride solution using polyetherimide/graphene composites. The Canadian Journal of Chemical Engineering 94: 896–904.

[40] Yu, F., L. Camilli, T. Wang, D. M. A. Mackenzie, M. Curioni, R. Akid and P. Bøggild. 2018. Complete long-term corrosion protection with chemical vapor deposited graphene. Carbon 132: 78–84.

[41] Zheng, H., M. Guo, Y. Shao, Y. Wang, B. Liu and G. Meng. 2018. Graphene oxide–poly (urea–formaldehyde) composites for corrosion protection of mild steel. Corrosion Science 139: 1–12.

[42] Irfan, M., S. I. Bhat and S. Ahmad. 2018. Reduced graphene oxide reinforced waterborne soy alkyd nanocomposites: formulation, characterization, and corrosion inhibition analysis. ACS Sustainable Chemistry & Engineering 6: 14820–14830.

[43] Surudžić, R., A. Janković, N. Bibić, M. Vukašinović-Sekulić, A. Perić-Grujić, V. Mišković-Stanković, S. J. Park and K. Y. Rhee. 2016. Physico–chemical and mechanical properties and antibacterial activity of silver/poly (vinyl alcohol)/graphene nanocomposites obtained by electrochemical method. Composites Part B: Engineering 85: 102–112.

CHAPTER 8

Metal-Graphene Nanocomposites with Improved Mechanical and Anti-Corrosion Properties

Liana Maria Muresan

1. General Aspects

Metal-graphene nanocomposites are relatively new, but very promising players in the smart materials market, mainly due to the exceptional properties of graphene, a true 2D material: high surface area, excellent conductivity, high mechanical strength, ease of functionalization and mass production [1]. The metal-graphene composites are versatile, eco-friendly materials that can be used in many domains such as green energy production in fuel cells and microbial cells, detection of chemical species and are viable alternatives to conventional anticorrosive coatings [2].

There are several forms in which graphene can be introduced in the metallic matrix: as graphene (Gr), graphene oxide (GO), or reduced graphene oxide (rGO). Graphene (Gr) is a sp^2 hybridized monolayer of carbon atoms that are bonded together in a hexagonal honeycomb lattice [3]. Layers of graphene stacked on top of each other form graphite.

Graphene oxide (GO) is a single-atomic layered material made by the exfoliation of graphite, which is a cheap and accessible raw material. GO is not a good conductor, but one of its main advantages is that it can be easily dispersed in water and organic solvents, as well as in different matrices, due to the presence of the oxygen functionalities [4, 5]. Moreover, graphene oxide is cheaper and easier to prepare than graphene and can be easily converted to graphene by various methods such as thermal microwave synthesis [6], chemical [7] or electrochemical methods [8–11], etc.

The reduced graphene oxide (rGO) has many structural defects and functional groups which are advantageous for electrochemical applications such as photovoltaic

Department of Chemical Engineering, "Babes-Bolyai" University, 11, Arany Janos St. 400028 Cluj-Napoca, Romania.
Email: lianamuresan2002@yahoo.com

cells and supercapacitors, electronic devices, etc. Moreover, there are ways to functionalize rGO even more by various chemical and electrochemical methods in order to enhance its efficiency [12, 13]. Graphene functionalization can significantly improve some of the composite's characteristics (dispersion degree in the metallic matrix, interaction with biomolecules, etc.) leading to high-performing systems. Due to the increased interest in composites embedding graphene, in this chapter some obtaining methods for various metal/alloys-graphene composites are reviewed. The influence of different experimental factors on the anti-corrosion performance of these materials is also analyzed. Additionally, the potential of most common metal-graphene composites as corrosion protective layers on metals and alloys is examined and some working mechanisms proposed in the literature by different authors are discussed. Other applications of metal-graphene composites are also briefly reviewed.

2. Preparation Methods of Metal-Graphene Composites

The key factor in obtaining metal-graphene composites with excellent properties is the uniformity of the deposits. There are many challenges involved to get uniformly dispersed graphene in metal matrix composites, due to the differences between their densities and those of the metals in which they should be incorporated. This is why different preparation methods are described in the literature for metal-graphene composites.

2.1 Electrodeposition

Electrodeposition is one of the most widely used methods for metal-graphene composites preparation, due to its advantages: (i) low temperature operation (which preserves the properties of graphene during the process) (ii) lack of dangerous reactants (iii) low cost and (iv) scalability [3]. The hydrophilic graphene oxide (GO) can be easily dispersed in polar solvents, which makes it a facile material to fabricate graphene-based coatings by the electrochemical technology [14]. Moreover, electrodeposition of metals in the same time with graphene oxide reduction ensures a more uniform distribution of graphene into the metallic coating and hence, better mechanical and anticorrosion properties of the resulting composite material.

In principle, electrodeposition can be achieved by using either DC [15–19] or pulsed current [20–22] in a plating electrolyte solution containing dispersed graphene. The electroplating processes for Ni, Cu and Zn are well established and have been already used to obtain metal-graphene composites with improved mechanical and corrosion resistant properties.

Among other experimental factors, the zeta potential of graphene in the plating bath plays an important role in its incorporation in the cathodic deposit. For example, in a zinc plating bath containing $ZnSO_4$, Na_2SO_4, NaCl, CTAB and graphene, the zeta potential of graphene was found to be + 11.8 mV [16]. This positive charge ensures the migration of the particles toward the cathode and their incorporation in the zinc matrix.

The stability of graphene suspensions is another factor that should be taken into consideration and ultrasonication is frequently used along with the use of surfactants.

Decoration of graphene with oxygenated functional groups enhancing hydrophilicity could be also considered as a possibility to improve dispersability [23]. The uniform dispersion of graphene in the metallic matrix is difficult to achieve and remains a challenge. The reduction of GO is a key topic, as different reduction methods lead to different properties of materials containing rGO. Functional groups and defects can significantly influence the conductivity of GO and its reduction easiness. The concentration of lattice defects in the carbon plane is the key for a complete reduction of GO.

The reduction process of the GO film can be controlled by the choice of pH [24], reduction potential and electrolyte medium [10]. In the case of GO voltammetric reduction, the reduction begins at –0.6 V and reaches a maximum at –0.87 V, being irreversible [23]. On the other hand, a low pH is favorable to GO reduction, suggesting that H^+ ions are involved in the reduction process [24].

2.2 Electrophoresis

Deposition of graphene on a metallic substrate from a colloidal suspension using electrophoresis (EPD) is a promising method to develop anti-corrosion protective coatings. In principle, during electrophoresis, electrostatically charged colloidal particles are deposited onto an oppositely charged electrode by the application of a DC voltage resulting in coatings with controlled structure and thickness.

The main advantages of this technique are (i) the work at room temperatures, (ii) better control of coating thickness and (iii) high deposition rate [25]. However, due to the poor dispersion of the graphene particles in aqueous solutions, its direct deposition could lead to unsatisfactory results. On the contrary, graphene oxide can be easily dispersed in water by ultrasonication, due to the functional groups grafted on its surface. Consequently, GO deposition on metals is preferred either in pure state or together with a polymer [26]. The deposition of GO by electrophoresis results in partial reduction of GO. For example, by depositing GO on copper anode, the carboxylate groups present on GO release CO_2 by Kolbe reaction and the resulted graphene form covalent bonds with GO, thus consolidating the deposited layer. The experimental conditions reported for GO suspension electrophoretic deposition on copper were: DC voltage 2–15 V, inter-electrode gap 3 cm, duration 10s [25]. In other cases, the optimal EPD conditions were DC voltage 10 V, inter-electrode gap 1 cm, deposition time 30s for a deposit thickness of 40 nm [27]. A polymeric isocyanate cross-linked with hydroxyl functional acrylic adhesive (PIHA) was used as polymer matrix for GO nanoplatelets deposited electrophoretically. As expected, the thickness of the coating increased with the increase of voltage. To reduce or eliminate the coating porosity, a heat treatment should be applied after the deposition.

To conclude, EPD is a cost effective, versatile and fast method, offering the possibility to control the microstructure, dimensions and uniformity of the resulting coating.

2.3 Electroless Plating

Electroless plating is another method to obtain metal graphene composites. Iron-graphene coatings on copper substrates were prepared from an alkaline plating bath in the absence of reducing agents by connecting the copper with aluminum [28]. A galvanic cell is formed between the two metals in which aluminum plays the role of electron source for iron deposition, the electrons being transferred to the copper substrate and further to Fe^{2+} ions. GO in suspension might be reduced by Fe^{2+} ions as well as by the hydrogen generated in the reaction between Al and OH^- ions from the alkaline plating bath and is incorporated in the Fe matrix. The plating process was carried out at 40°C, for 60 min. The Fe-rGO composite coating on copper prepared from the reducing agent-free plating bath possessed better corrosion resistance than pure Fe coatings.

Nickel phosphorous can also be co-deposited with rGO by using either *in situ* or *ex situ* electroless deposition [29] from a solution containing nickel chloride, sodium citrate, sodium hypophosphite and cetyl trimethyl ammonium bromide. The hardness of nickel-phosphorus-rGO coating was found to be much lower than that of the electroless nickel-phosphorus deposits, suggesting that the graphene was behaving as soft material reinforcement.

Electroless plating method was used to prepare Ni-decorated graphene as an enhancing component in a copper matrix. The electroless plated graphene-copper composite bulk has a better performance in the mechanical tensile strength compared to pure copper, and also maintains superior ductility, elongation and electrical conductivity [30]. It was suggested that Ni enhances the interfacial compatibility between graphene and copper, ensuring homogeneous dispersion of graphene sheets in the copper matrix.

2.4 Chemical Vapor Deposition

In chemical vapor deposition, (CVD), the metallic substrate is exposed to one or more volatile precursors, which react and/or decompose on the substrate surface to produce the desired deposit. CVD was successfully used to deposit 3D flexible and conductive interconnected graphene networks on porous Ni surfaces used as anodes in microbial fuel cells [31]. The graphene coating was proven to be resistant to microbial corrosion, by acting as a barrier that prevents the access of both microbes and microbial byproducts to the Ni surface.

It was reported that CVD growth of graphene on copper films requires relatively thick films (~ 1 μm thick) as-deposited metal thin films are unstable and 30% of thin film can evaporate during the growth after heating at 1000°C. To avoid this situation, very thin metal catalyst films (e.g., Ni-Cu alloy) can be used to promote monolayer growth thus improving the monolayer fraction of graphene coverage on the thinner films [32]. Multilayers of metal-graphene composites were obtained by alternating layers of metal (copper or nickel) and monolayered graphene [33].

2.5 Other Methods

In order to obtain uniformly distributed GO inside the metallic matrix, a binder such as polyvinyl alcohol (PVA) can be used [34] during *flake powder metallurgy* (Flake PM) to obtain Al matrix composites. The name Flake PM was derived from the use of flake metal powders, resulting in the uniform dispersion of reinforcements in the metal matrices and thus in balanced strength and ductility. The use of PVA has some floods, as an incomplete pyrolysis of PVA leads to the appearance of residual impurities in the Al-GO composite. To avoid these inconveniences, the negatively charged GO sheets in aqueous solutions are forced to interact with the ionized Al flakes which are transformed in Al^{3+} ions due to their negative standard reduction potential value. Electrostatic interaction offers a possibility to develop high-performance graphene reinforced metal matrix composites. Taking into account that between reduced graphene oxide (rGO) and Al flakes spontaneous electrostatic interactions occur, uniform distribution of rGO in an Al matrix can be achieved [35, 36]. After annealing, which leads to GO reduction to graphene, uniformly distributed GO in the Al matrix is accomplished.

Thin films of graphene on metals can be obtained by using *doctor blade technique* (or tape casting). This technique is widely used for producing thin films on large area surfaces starting from mixtures of a suspension of ceramic particles along with additives (such as dispersants, binders, or plasticizers). The mixture is placed on a substrate beyond a doctor blade and a constant relative movement is established between the blade and the substrate. Thus, the slurry spreads on the substrate to form a thin film which results in a gel-layer upon drying [37]. Mg alloy AZ91 coated with a thin film (~ 1 μm thickness) of electrochemically produced graphene by using doctor blade technique exhibited increased corrosion resistance in NaCl, KCl and Na_2SO_4 solutions [38].

Mechanical mixing of graphene with metals (e.g., Al) by using ball milling under inert atmosphere prevents particles agglomeration and allows the obtaining of bulk metal composites with fine grains [39]. Graphene nanoplatelets (GNPs) were *mechanically impregnated* into aluminum substrates to form surface nanocomposites, which enhance the surface properties of Al. The impregnation was achieved by application of pressure on graphene coated aluminum plates which are locally softened by electrical resistance heating [40].

3. Properties of Metal-Graphene Composites

As any other composite material, metal-graphene composites possess different physical and chemical properties than the starting materials from which they were prepared. Some of these properties are presented in what follows.

3.1 Morphology and Structure

A change of the surface morphologies and microstructure of the metal-graphene deposits is visible in almost all cases. A decrease of grain size [16, 28], a change of preferential orientation [15, 29], smaller nodules rise and denser structure [28] can be observed for various metal-graphene combinations. The strong interaction between the metal (e.g., Ni) and Gr can also be confirmed from the increase in interplanar spacing parameter with the incorporation of Gr [15].

As expected, the grain size of the deposits is influenced by the experimental conditions during the obtaining process. Thus, it was reported that the grain size of the Ni-Gr composites synthesized by electrodeposition with a lower current density is smaller than that with a higher current density, and the roughness of the surface of the composites is smaller at a low current density than that at a high current density [15]. The tendency of Ni to grow along <111> instead of <200> direction could be attributed to a growth of the metal on the graphene sheets during the electrodeposition process [41].

In the case of Cu-Gr coatings, the incorporation of the graphene in the Cu matrix enhanced the crystal growth along the <200> and <220> directions and suppressed growth along the <111> direction, which was characteristic in the case of pure Cu [18]. The tendency was enhanced by the increase of incorporated graphene in the composite coating. It was speculated that the change of the preferential growth is due to a change of the surface energy of the metal during electrodeposition. The modification of the morpho-structural characteristics explains the improvement of mechanical properties and of the corrosion resistance of metal-graphene composites.

3.2 Mechanical Properties

Investigation of the mechanical properties of metal-graphene composites has been receiving increasing attention due to their high elastic modulus, increased hardness and improved wear resistance. It is remarkable that metal-graphene nanocomposites with low volume/weight fractions of graphene inclusions exhibit dramatically enhanced strength and hardness. The key effects of graphene nanosheets and nanoplatelets on mechanical characteristics of metal-graphene nanocomposites are: (i) dislocation strengthening, (ii) stress transfer and (iii) grain refinement of metallic matrices [42].

A uniform dispersion of graphene inside the metallic matrix is a condition for a beneficial reinforcement effect. Thus, the reinforcing efficiency of rGO in rGO/ Al composites is one order of magnitude higher than the conventional microscale reinforcement of Al matrix composites [35].

It was also reported that the incorporation of graphene nanosheets into a Ni matrix increases the elastic modulus and hardness of Ni-Gr composite to values which are 1.2–1.7 times higher than that of pure Ni deposited under the same conditions [15]. The increase of elastic modulus could be due to the orientation degree of crystallites in the <111> direction, knowing that <111> plane possesses higher elastic modulus than other planes. The microhardness increases with the graphene content in the electrolyte used in electrodeposition, mostly due to grain refining of the deposit [21].

The average friction coefficient and wear rates of the metal-graphene composite coatings are also different from those of the pure metallic coatings. The results confirm that in the case of Co-GO composites, the friction is reduced, and the wear resistance is improved due to the self-lubricating properties of GO nanosheets [43]. The wear surface of Co-GO coating is featured with slight plastic deformation, which explains its good mechanic resistance. The same increase in the wear resistance and decrease of the friction coefficient were noticed in the case of Ni-Gr composites [21]. Improved mechanical properties were reported also in the case of Cu-Gr composites [22, 44].

3.3 Corrosion Resistance

One of the most valuable properties acquired by the metal-graphene deposits consists in better corrosion resistance than that of the pure metals and alloys. Graphene is highly inert and so can act as a corrosion barrier between oxygen and water diffusion when present on metallic surfaces. On the other hand, the presence of graphene in metal matrix composites enhances the corrosion resistance of the materials by altering the morphology and texture of the deposits. For example, in the case of Fe-rGO coatings prepared on copper substrates using an electroless plating method, Fe-rGO coating had the smaller nodule size and denser structure owing to the GO added in the plating bath, and therefore possessed better corrosion resistance as compared to Fe coatings [28]. Improved corrosion resistance was reported also in the case of Mg-Gr [38], Zn-Gr [16], Cu-GO even in highly corrosive media such as those containing Cl⁻ ions [17, 18].

A possible mechanism for enhanced corrosion resistance of Cu-GO coatings in Cl⁻ containing media was proposed by Raghupathy et al. [18]. This mechanism essentially relates to texture induced passivity at the local sites of passivity breakdown, being only valid within the defect sites in the passive film, where Cu corrosion is most likely to occur. The mechanism is based on three stages: (i) the pre-existence of a passive film on Cu and Cu-GO surface with some defects that could become preferential sites for corrosion; (ii) anodic dissolution of Cu at the defect sites and (iii) formation of $CuCl_2^-$ complexes inside defects followed by precipitation of CuO_2 film within the defect which restores the passivity. Similar results, consistent with 6–10 times lower corrosion current densities at GO-modified surface than that at bare copper, were reported for Cu-GO coatings prepared by electrophoretic deposition [25, 27].

Ni-graphene composite was proven to be resistant also to microbial corrosion [31]. The inert graphene coating deposited on Ni foam via chemical vapor deposition prevents the access of microbes to the metallic surface by acting as a protective barrier. In the same time, the coating hinders the transport of Ni^{2+} ions at metal/solution interface and protects the Ni surface from microbial byproducts that favor Ni dissolution.

Electroless nickel phosphorous coatings containing reduced graphene oxide (rGO) deposited on mild steel have better corrosion resistance than the same coatings without graphene. rGO prevents Cl⁻ ion attack and reduces the internal stress of the coating [29].

The methods currently used to investigate the electrochemical corrosion of composite metal-graphene coatings are linear and potentiodynamic polarization (Tafel), and electrochemical impedance spectroscopy. These methods make possible the determination of kinetic parameters associated with the corrosion process and offer indications related to the corrosion mechanism. Better anti-corrosion properties are reflected in increased values of the polarization resistance, smaller values of corrosion current density and, in some cases, in lower double layer capacitance. An example of potentiodynamic polarization curves for Zn and Zn-electroreduced GO coatings on steel is presented in Fig. 1.

It can be observed that the presence of graphene influences both the anodic and cathodic branches of the polarization curves, suggesting that on one side, the

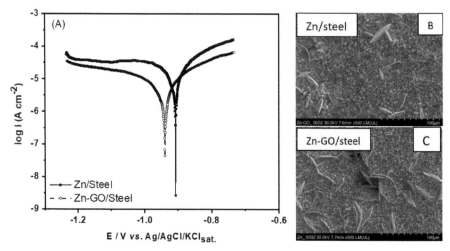

Fig. 1. Tafel polarization curves of Zn and Zn-electrolytically reduced GO composite coated steel (A) and the corresponding SEM micrographs (B and C, respectively).

coating acts as a barrier that impedes the diffusion of dissolved oxygen towards the metallic surface and, on the other side, hinders Zn dissolution. The changes in the coating structure and morphology put on evidence by SEM micrographs could be responsible for the better corrosion resistance of the composite deposit.

3.4 Others

The thermal conductivity of metal-graphene composites is strongly anisotropic and enhanced in the direction perpendicular to the layers. It was reported that the volume fraction of carbon, its spatial distribution and carbon-matrix interphase properties affect the macroscopic thermal conductivity of graphene-metal composites [45].

Graphene has high thermal and electrical conductivities, which are very useful characteristics for reinforcement of Cu-Gr composites without damaging the conductive properties of copper [46]. To obtain good electrical and thermal properties of the composite, the existence of hydroxyl groups on graphene surface should be avoided and an effective way to reach this purpose is to use electrodeposition assisted by pulse supply and ultrasonic stir [22]. The aggregation of Gr inside the composite should be also avoided because it can lead to increase in resistivity. The thermal conductivity of Ni-graphene composites prepared by electrodeposition was reported to be about 15% higher than that of pure Ni [41].

Soft magnetic films with improved characteristics and high frequency properties, such as FeCo-rGO coatings on ITO-glass, can be obtained by one step electrodeposition method [47]. The presence of graphene in the electrolyte increases the atomic Fe/Co ratio, suggesting that the deposition speed of the two metals is influenced by the graphene, possibly due to the chemical bond between the GO and Fe or Co ions. The magnetic anisotropy field and saturation magnetization of FeCo-RGO composite film are higher compared to pure FeCo alloy.

Some metal-graphene composites exhibit excellent catalytic properties that can be exploited for practical purposes. Thus, Al or Fe-graphene oxide composite (GO/metal composite) can be used as a novel heterogeneous acid catalyst that can be employed for the esterification of stearic acid and the reduction of high free fatty acid level of *Calophyllum inophyllum* oil [48]. Nanocomposites of pristine graphene with palladium (Pd-G) using swollen liquid crystals exhibited exceptionally better catalytic activity compared with Pd-RGO nanocomposite in the hydrogenation of nitrophenols and microwave assisted C-C coupling reactions [49].

Some of the most common metal-graphene composites, their obtaining methods and their properties are presented in Table 1.

Table 1. Summary of some metal-graphene composites and their properties.

No.	Composite	Obtaining method	Properties	Reference
1	Al-rGO	Electrostatic interaction	- improved elastic modulus and hardness	[35]
2	Ni-Gr	Electrodeposition on mild steel	- increase of the degree of preferential orientation - improved elastic modulus and hardness	[15]
3	Ni-Gr	Electrodeposition from sulfate bath on steel	- grain refining of the deposit - changed preferential orientation of crystals - increased hardness - higher corrosion resistance	[50]
4	Ni-Gr	Electrodeposition from sulfamate bath	- improved thermal conductivity - increased hardness	[41]
5	Ni-Gr	Pulse electrodeposition from Watts electrolyte	- increase of microhardness and wear resistance - decrease of friction coefficient	[21]
6	Ni-P-rGO	Electroless deposition on steel	- change in preferred orientation of Ni - better corrosion resistance	[29]
7	Ni-Gr	Electroless deposition	- enhanced tensile strength - good ductility, elongation and electrical conductivity	[30]
8	Ni_{foam}-Gr	Chemical vapor deposition (CVD)	- increased protection against microbial corrosion	[31]
9	Co-Gr	Pulse electrodeposition from sulfate bath	- improved tribological properties - enhanced wear and corrosion resistance	[43]
10	FeCo-rGO	One step electro-deposition on ITO glass	- soft magnetic properties - increased anisotropy and saturation magnetization - improved ferromagnetic resonance frequency	[45]
11	Fe-rGO	Electroless plating on Cu connected with Al foils from alkaline plating bath	- smaller nodules rise and denser structure - better corrosion resistance	[28]

Table 1 Contd. ...

...Table 1 Contd.

No.	Composite	Obtaining method	Properties	Reference
12	Cu-Gr	Pulse electrodeposition from sulfate bath with ultrasound stirring	- improved mechanical properties - same thermal and electrical properties with pure copper	[22]
13	Cu-GO	Electrodeposition on steel	- improved corrosion resistance	[17]
14	Cu-rGO	Electrophoretic deposition on copper	- better corrosion resistance even in highly corrosive media (Cl-)	[27]
15	Cu-Gr	Electrodeposition on steel	- enhanced corrosion resistance - reduction of crystallite size - strong <220> texture	[17,18]
16	Cu-GO	Electrophoresys	- increased corrosion resistance	[25]
17	Cu-graphene	Chemical vapor deposition (CVD)	- different surface roughening behavior of copper	[51]
18	Cu-Gr	microwave and conventionally sintered processes	- superior electrical and mechanical properties than conventional counterparts	[52]
19	Sn-Gr	Electrodeposition	- changes in morphology, grain size and texture - decrease in corrosion rate	[19]
20	Zn-Gr	Electrodeposition on steel	- microstructural changes (grain refining) - enhanced corrosion resistance	[16]
21	Mg-Gr	Blade coating technique	- improved corrosion resistance	[38]

4. Applications of Metal-Graphene Composites

There are numerous motives to develop metal-graphene composites. Due to the exceptional properties of graphene, which can be transferred to the composites incorporating them, several applications have been developed, varying from solar and fuel cells, to sensors and protective coatings. Besides their use as anticorrosion coatings, some of them have useful applications that are mentioned below.

4.1 Sensors and Biosensors

Graphene based nanocomposites have been developed as an enhanced sensing platform for sensors/biosensors. Because their 2D structure provides a large area for the immobilization of enzymes and because their enriched oxygen-containing groups are able to react with amino-acid residues, graphene derivatives are ideal substrates for immobilization of certain biomolecules, through electrostatic interactions, covalent linkage or entrapment in a polymeric matrix [53].

As an example, electrodeposition was employed to prepare Au nanostructured film on a graphene-HRP-chitosan modified electrode. The resulting biosensor exhibits excellent electrocatalytic response to H_2O_2, along with a wide linear range and long-term stability [54]. Cyclic voltammograms showed that the biosensor realized the direct electron transfer between HRP and the electrode and exhibited the typical catalytic reduction of H_2O_2. A similar biosensor was realized by layer-by-layer assembly of hemoglobin, gold nanoparticles, chitosan and graphene onto glassy carbon electrode [55] and was successfully used for H_2O_2 detection. Nanostructures made of titanium dioxide and/or platinum nanoparticles supported on reduced graphene oxide (RGO) were synthesized by a simple and fast microwave-assisted route to develop a sensor for H_2O_2 detection [56].

4.2 Energy Conversion

Graphene-based materials in different forms of 1D, 2D to 3D have proven to be excellent candidates of electrode materials in electrochemical energy storage systems [57].

In order to improve the performance of rechargeable lithium-ion batteries (LIBs) carbonaceous materials, metal/metal oxide, metal sulfide and metal-free elements can be integrated into 3D graphene network to form composites for LIBs cathodes [58]. 3D porous graphene hybrids exhibit significantly improved reversible capacity, rate capability, and desirable cycling performances. Multi-layer graphene was synthesized on a nickel foam template by chemical vapor deposition and loaded with nickel oxide nanostructures using the successive ionic layer adsorption and reaction technique. The resulting three-dimensional composites were characterized and investigated as electrode material for supercapacitors, which are considered valuable alternatives to other energy storage devices [59]. Incorporating pseudo-active materials such as metals, metal oxides, polymers, etc. into 3D graphene architectures is one of the most promising ways to improve supercapacitors' performance [58]. Graphene/metal oxides or hydroxides composites are promising materials for achieving both power density and high energy density required by energy storage devices [57].

Considerable efforts have been dedicated to the design of fuel cells with outstanding energy density, acceptable operating temperatures and low impact for environment. With high porosity and multidimensional electron transport path, 3D graphene structures are highly desirable for catalyst loading in fuel cells. These structures facilitate the mass transfer and maximize the accessibility of the electroactive species to the catalyst surfaces [60]. Thus, a thermally reduced graphene oxide grown with carbon nanotubes composite was used as 3D highly conductive carbon scaffolds, where a large amount of small and homogeneous Pt nanoparticles (from 3.37 ± 1.22 to 4.24 ± 1.83 nm) was directly synthesized to acquire a new type of catalyst used in methanol fuel cells as anode material [61]. The oxidation of methanol on the Pt/3D graphene was more efficient than that of commercial Pt/carbon black and Pt/rGO sheets.

Graphene–metal particle nanocomposites were prepared in a water–ethylene glycol system using graphene oxide as a precursor and metal nanoparticles (Au, Pt and Pd) as building blocks [2]. The composite material proved its potential application in direct methanol fuel cells.

Metal/graphene has been used also as a filter membrane exterior to the hydrogen fuel cell to prevent CO poisoning [62].

4.3 Biomedical Applications

One of the promising developments of metal-graphene composites is their use in biomedical applications for diagnosis, therapy and sensing. Uniform and water-soluble Ag-reduced graphene oxide (Ag-rGO) nanocomposites display much better antibacterial properties than pure silver nanoparticles and an equivalent antibacterial effect in comparison with that of the general antibacterial drug ampicillin [63].

Different transitional metals can be used to decorate graphene to produce nanoclusters (NCs) that are included in sensing devices for biomedical applications [64]. NCs are clusters of nanoparticles with sizes of 1–20 nm, with narrow size distribution and with good photostability. They can be stabilized with proteins and used for targeted sensing of cancer cells. An example could be the sensor obtained by immobilization of peptide-stabilized AuNCs on GO [65]. The peptide-gold nanocluster/graphene nanocomplex yielded an intense "turn-on" fluorescent response, which is strongly correlated with the enzyme concentration. The sensor was successfully applied for detection of metalloproteinase-9 secreted from human breast adenocarcinoma MCF-7 cells with high sensitivity, selectivity, significant improvement in terms of detection time and simplicity. A novel label-free fluorescent biosensor platform has been developed for protease activity assay using peptide-templated gold nanoclusters (AuNCs) [66]. A recent finding is that enzymes are able to exert chemical modifications on the peptide-templated AuNCs and quench their fluorescence, which furnishes the development of a real-time and label-free sensing strategy for post-translational modification (PTM) enzymes [67].

Besides cancer cells detection, metal-graphene nanoclusters (e.g., nucleic acids stabilized AgNCs immobilized on GO) can be also used for pathogenic DNA sensing, specifically for certain genes detection [20]. This research domain is very promising, and works are in progress. A glucose non-enzymatic biosensor working in alkaline medium was obtained by simultaneous reduction/deposition of reduced graphene oxide/copper nanoparticles (rGO/Cu NPs) on a glass/Ti/Au electrode using electrophoretic deposition (EPD) technique [68]. The interferences from various oxidizable molecules such as dopamine, uric acid, ascorbic acid and carbohydrate molecules such as fructose, lactose and galactose were negligible.

5. Conclusions and Future Perspectives

Metal-graphene composites are promising materials for the fabrication of nanostructured polymers, super-capacitor devices, drug delivery systems, solar cells, memory devices, transistor devices, biosensors, etc. They can be prepared either by dispersing the graphene in a metallic matrix or by forming layered structures with alternative layers of metal and graphene. There are a lot of preparation methods for metal-graphene composites, including mechanical mixing, electrodeposition, electrophoresis, chemical vapor deposition, etc., each with its advantages and disadvantages.

The resulting composites exhibit improved mechanical, anticorrosive and catalytic properties, which entitles them as valuable alternatives to conventional materials for applications in many fields, such as automobile and aerospace industries, energy conversion, biomedicine, nanoelectronics, and many others. The key towards the successful application of graphene lies in their modification and/or integration into high quality composite nanomaterials. Systematic studies on metal-graphene properties and mechanisms of action will certainly lead to a tremendous development of their applicability.

References

[1] Shao, Y., J. Wang, H. Wu, J. Liu, I. A. Aksay and Y. Lin. 2010. Graphene based electrochemical sensors and biosensors: a review. Electroanalysis 22: 1027–1036.

[2] Xu, C., X. Wang and J. Zhu. 2008. Graphene–metal particle nanocomposites. J. Phys. Chem. C 112: 19841–19845.

[3] Xavior, M. A. and H. G. Prashantha Kumar. 2017. Processing and characterization techniques of graphene reinforced metal matrix composites: A review. Mater. Today Proc. 4: 3334–3341.

[4] Paredes, J. I., Rodil, M. J. F. Merino, L. Guardia, A. M. Alonso and J. M. D. Tarascon. 2011. Environmentally friendly approaches toward the mass production of processable graphene from graphite oxide. J. Mater. Chem. 21: 298–306.

[5] Konios, D., M. M. Stylianakis, E. Stratakis and E. Kymakis. 2014. Dispersion behaviour of graphene oxide and reduced graphene oxide. J. Coll. Interf. Sci. 430: 108–112.

[6] Huang, W., Q. Hao, W. Lei, L. Wu and X. Xia. 2014. Polypyrrole-hemin-reduce graphene oxide: rapid synthesis and enhanced electrocatalytic activity towards the reduction of hydrogen peroxide. Mater. Res. Express 1: 045601.

[7] Stankovich, S., D. A. Dikin, R. D. Piner, K. A. Kohlhaas, A. Kleinhammes, Y. Jia et al. 2007. Synthesis of graphene-based nanosheets via chemical reduction of exfoliated graphite oxide. Carbon 45: 1558–1565.

[8] Low, C. T. J., F. C. Walsh, M. H. Chakrabarti, M. A. Hashim and M. A. Hussain. 2013. Electrochemical approaches to the production of graphene flakes and their potential applications. Carbon 54: 1–21.

[9] Deng, K. Q., J. Zhou and X. F. Li. 2013. Direct electrochemical reduction of graphene oxide and its application to determination of l-tryptophan and l-tyrosine. Colloids Surf. B: Biointerfaces 101: 183–188.

[10] Kauppila, J., P. Kunnas, P. Damlin, A. Viinikanoja and C. Kvarnström. 2013. Electrochemical reduction of graphene oxide films in aqueous and organic solutions. Electrochim. Acta 89: 84–89.

[11] Lindfors, T., A. Österholm, J. Kauppila and M. Pesonen. 2013. Electrochemical reduction of graphene oxide in electrically conducting poly(3,4-ethylenedioxythiophene) composite films. Electrochim. Acta 110: 428–436.

[12] Kuila, T., S. Bose, A. K. Mishra, P. Khanra, N. H. Kim and J. H. Lee. 2012. Chemical functionalization of graphene and its applications. Prog. Mater Sci. 57: 1061–1105.

[13] Chakrabarti, M. H., C. T. J. Low, N. P. Brandon, V. Yufit, M. A. Hashim, M. F. Irfan et al. 2013. Progress in the electrochemical modification of graphene-based materials and their applications. Electrochim. Acta 107: 425–440.

[14] Qiu, C., D. Liu, K. Jin, L. Fang and T. Sha. 2017. Corrosion resistance and micro-tribological properties of nickel hydroxide graphene oxide composite coating. Diam. Relat. Mater. 76: 150–156.

[15] Ren, Z., N. Meng, K. Shehzad, Y. Xu, S. Qu, B. Yu et al. 2015. Mechanical properties of nickel-graphene composites synthesized by electrochemical deposition. Nanotechnol. 26: 065706–065713.

[16] Punith Kumar, M. K., M. Pratap Singh and C. Srivastava. 2015. Electrochemical behavior of Zn–graphene composite coatings. RSC Adv. 5: 25603–25608.

[17] Raghupathy, Y., A. Kamboj, M. Y. Rekha, N. P. Narasimha Rao and C. Srivastava. 2017. Copper-graphene oxide composite coatings for corrosion protection of mild steel in 3.5% NaCl. Thin Solid Films 636: 107–115

[18] Kamboj, A., Y. Raghupathy, M. Y. Rekha and C. Srivastava. 2017. Morphology, texture and corrosion behavior of nanocrystalline copper–graphene composite coatings. JOM. DOI. 10.1007/s11837-017-2364-0.

[19] Berlia, R., M. K. Punith Kumar and C. Srivastava. 2015. Electrochemical behavior of Sn–graphene composite coating. RSC Adv. 5: 71413–71418.

[20] Liu, X., F. Wang, R. Aizen, O. Yehezkeli and I. Willner. 2013. Graphene oxide/nucleic-acid-stabilized silver nanoclusters: functional hybrid materials for optical aptamer sensing and multiplexed analysis of pathogenic DNAs. J. Am. Chem. Soc. 135: 11832–11839.

[21] Algul, H., M. Tokur, S. Ozcan, M. Uysal, T. Cetinkaya, H. Akbulut et al. 2015. The effect of graphene content and sliding speed on the wear mechanism of nickel–graphene nanocomposites. Appl. Surf. Sci. 359: 340–348.

[22] Huang, G., H. Wang, P. Cheng, H. Wang, B. Sun, S. Sun et al. 2016. Preparation and characterization of the graphene-Cu composite film by electrodeposition process. Microelectron. Eng. 157: 7–12.

[23] Pei, S. and H. -M. Cheng. 2012. The reduction of graphene oxide. Carbon 50: 3210–3228.

[24] Zhou, M., Y Wang, Y Zhai, J Zhai, W Ren, F Wang et al. 2009. Controlled synthesis of large-area and patterned electrochemically reduced graphene oxide films. Chem. Euro J. 15: 6116–6120.

[25] Raza, M. A., Z. U. Rehman, F. A. Ghauri, A. Ahmad, R. Ahmad and M. Raffi. 2016. Corrosion study of electrophoretically deposited graphene oxide coatings on copper metal. Thin Solid Films 620: 150–159.

[26] Singh, B. P., B. K. Jena, S. Bhattarchajee and L. Besra. 2013. Development of oxidation and corrosion resistant hydrophobic graphene oxide-polymer composite coating on copper. Surf. Coat. Technol. 232: 475–481.

[27] Singh, B. P., S. Nayak, K. K. Nanda, B. K. Jena, S. Bhattacharjee and L. Besra. 2013. The production of a corrosion resistant graphene reinforced composite coating on copper by electrophoretic deposition. Carbon 61: 47–56.

[28] Zhang, X., Y. Zhou, A. Liang, B. Zhang and J. Zhang. 2016. Facile fabrication and corrosion behavior of iron and iron-reduced graphene oxide composite coatings by electroless plating from baths containing no reducing agent. Surf. Coat. Technol. 304: 519–524.

[29] Sadhir, M. H., M. Saranya, M. Aravind, A. Srinivasan, A. Siddharthan and N. Rajendran. 2014. Comparison of *in situ* and *ex situ* reduced graphene oxide reinforced electroless nickel phosphorus nanocomposite coating. Appl. Surf. Sci. 320: 171–176.

[30] Jiang, R., X. Zhou and Z. Liu. 2017. Electroless Ni-plated graphene for tensile strength enhancement of copper. Mater. Sci. Eng. A 679: 323–328.

[31] Krishnamurthy, A., V. Gadhamshetty, R. Mukherjee, Z. Chen, W. Ren, H. -M. Cheng et al. 2013. Passivation of microbial corrosion using a graphene coating. Carbon 56: 45–49.

[32] Cho, J. H., J. J. Gorman, S. R. Na and M. Cullinan. 2017. Growth of monolayer graphene on nanoscale copper-nickel alloy thin films. Carbon 115: 441–448.

[33] Kim, Y., J. Lee, M. S. Yeom, J. W. Shin, H. Kim, Y. Cui et al. 2013. Strengthening effect of single-atomic-layer graphene in metal–graphene nanolayered composites. Nat. Comm. 4: 2114.

[34] Wang, J. Y., G. Fan, Z. Tan, Q. Guo, D. Xiong, Y. Su et al. 2012. Reinforcement with graphene nanosheets in aluminum matrix composites. Scripta Mater. 66: 594–597.

[35] Li, Z., G. Fan, Z. Tan, Q. Guo, D. Xiong, Y. Su, Z. Li et al. 2014. Uniform dispersion of graphene oxide in aluminum powder by direct electrostatic adsorption for fabrication of graphene/aluminum composites. Nanotechnol. 25: 325601.

[36] Li, Z., G. Fan, Z. Tan, Z. Li, Q. Guo, D. Xiong and D. Zhang. 2016. A versatile method for uniform dispersion of nanocarbons in metal matrix based on electrostatic interactions. Nano-Micro Lett. 8: 54–60.

[37] Berni, A., M. Mennig and H. Schmidt. 2004. Doctor blade. pp. 89–92. *In*: Aegerter, A. M. and M. Menning (eds.). Sol-Gel Technologies for Glass Producers and Users. Springer Science & Business Media New York.

[38] Selvam, M., K. Saminathan, P. Siva, P. Saha and V. Rajendran. 2016. Corrosion behavior of Mg/graphene composite in aqueous electrolyte. Mater. Chem. Phys. 172: 129–136.

[39] Ebinezar, B. 2014. Analysis of hardness test for aluminium carbon nanotube metal matrix and graphene. Indian J. Eng. 10: 33–39.

[40] Sahoo, B., J. Joseph, A. Sharma and J. Paul. 2017. Surface modification of aluminium by graphene impregnation. Mater. Design 116: 51–64.

[41] Kuang, D., L. Xu, L. Liu, W. Hu and Y. Wu. 2013. Graphene–nickel composites. Appl. Surf. Sci. 273: 484–490.

[42] Ovid'ko, I. A. 2014. Metal-graphene nanocomposites with enhanced mechanical properties: a review. Rev. Adv. Mater. Sci. 38: 190–200.
[43] Liu, C., F. Su and J. Liang. 2015. Producing cobalt–graphene composite coating by pulse electrodeposition with excellent wear and corrosion resistance. Appl. Surf. Sci. 351: 889–896.
[44] Zhang, D. and Z. Zhan. 2016. Strengthening effect of graphene derivatives in copper matrix composites. J. Alloys Compd. 654: 226–233.
[45] Wejrzanowski, T., M. Grybczuk, M. Chmielewski, K. Pietrzak, K. J. Kurzydlowski and A. Strojny-Nedza. 2016. Thermal conductivity of metal-graphene composites. Mater. Design 99: 163–173.
[46] Gao, X., H. Yue, E. Guo, H. Zhang, X. Lin, L. Yao and B. Wang. 2016. Mechanical properties and thermal conductivity of graphene reinforced copper matrix composites. Powder Technol. 301: 601–607.
[47] Cao, D., H. Li, Z. Wang, J. Wei, J. Wang and Q. Liu. 2015. Synthesis, nanostructure and magnetic properties of FeCo-reduced graphene oxide composite films by one-step electrodeposition. Thin Solid Films 597: 1–6.
[48] Marso, T. M. M., C. S. Kalpage and M. Y. Udugala-Ganehenege. 2017. Metal modified graphene oxide composite catalyst for the production of biodiesel via pre-esterification of Calophyllum inophyllum oil. Fuel 199: 47–64.
[49] Vats, T., S. Dutt, R. Kumar and P. F. Siril. 2016. Facile synthesis of pristine graphene-palladium nanocomposites with extraordinary catalytic activities using swollen liquid crystals. Sci Rep. 6: 33053, 1–11.
[50] Praveen Kumar, C. M., T. V. Venkatesha and R. Shabadi. 2013. Preparation and corrosion behavior of Ni and Ni-graphene composite coatings. Mater. Res. Bull 48: 1477–1483.
[51] Tajima, N., T. Kaneko, J. Nara and T. Ohno. 2014. Carbon atom reactions in the initial stage of CVD graphene growth on copper: A first principles study. Jpn. J. Appl. Phys. 53: 05FD08-1.
[52] Ayyappadas, C., A. Muthuchamy, A. Raja Annamalai and D. K. Agrawal. 2017. An investigation on the effect of sintering mode on various properties of copper-graphene metal matrix composite. Adv. Powder Technol. 28: 1760–1768.
[53] Du, D., W. Zhang, A. M. Asiri and Y. Lin. 2014. Sensors based on carbon nanotube arrays and graphene for water monitoring. pp. 3–19. *In*: Anita Street, Richard Sustich, Jeremiah Duncan and Nora Savage (eds.). Nanotechnology Applications for Clean Water: Solutions for Improving Water Quality. Second Edition. Elsevier Inc.
[54] Zhou, K., Y. Zhu, X. Yang, J. Luo, C. Li and S. Luan. 2010. A novel hydrogen peroxide biosensor based on Au–graphene–HRP–chitosan biocomposites. Electrochim. Acta 55: 3055–3060.
[55] Zhang, L., G. Han, Y. Liu, J. Tang and W. Tang. 2014. Immobilizing haemoglobin on gold/graphene–chitosan nanocomposite as efficient hydrogen peroxide biosensor. Sens. Actuators B 197: 164–171.
[56] Leonardi, S. G., D. Aloisio, N. Donato, P. A. Russo, M. C. Ferro, N. Pinna et al. 2014. Amperometric sensing of H_2O_2 using Pt–TiO_2/reduced graphene oxide nanocomposites. ChemElectroChem. 1: 617–624.
[57] Ke, Q. and J. Wang. 2016. Graphene-based materials for supercapacitor electrodes—A review. J. Materiomics 2: 37–54.
[58] Wang, Q., Q. Wang, M. Li, S. Szunerits and R. Boukherroub. 2015. Preparation of reduced graphene oxide/Cu nanoparticle composites through electrophoretic deposition: application for nonenzymatic glucose sensing. RSC Adv. 5: 15861–15869.
[59] Bello, A., K. Makgopa, M. Fabiane, D. Dodoo-Ahrin, K. I. Ozoemena and N. Manyala. 2013. Chemical adsorption of NiO nanostructures on nickel foam-graphene for supercapacitor applications. J. Mater. Sci. 48: 6707–6712.
[60] Maiyalagan, T., X. Dong, P. Chen and X. Wang. 2012. Electrodeposited Pt on three-dimensional interconnected graphene as a free-standing electrode for fuel cell application. J. Mater. Chem. 22: 5286–5290.
[61] Jhan, J. -Y., Y. -W. Huang, C. -H. Hsu, H. Teng, D. Kuo and P. -L. Kuo. 2013. Three-dimensional network of graphene grown with carbon nanotubes as carbon support for fuel cells. Energy 53: 282–287.

[62] Li, K., Y. Li, H. Tang, M. Jiao, Y. Wang and Z. Wu. 2015. A density functional theory study on 3D metal/graphene for the removal of CO from H_2 feed gas in hydrogen fuel cells. RSC Adv. 5: 16394–16399.

[63] Xu, W. -P., L. -C. Zhang, J. -P. Li, Y. Lu, H. -H. Li, Y. -N. Ma et al. 2011. Facile synthesis of silver@ graphene oxide nanocomposites and their enhanced antibacterial properties. J. Mater. Chem. 21: 4593–4597.

[64] Muthoosami, K., R. G. Bai and S. Manickam. 2017. Graphene metal nanoclusters, in cutting edge theranostics nanomedicine applications. pp. 429–477. *In*: Anuj Tripathi and Jose Savio Melo (eds.). Advances in Biomaterials for Biomedical Applications. Springer.

[65] Nguyen, P. D., V. T. Cong, C. Baek and J. Min. 2017. Fabrication of peptide stabilized fluorescent gold nanocluster/graphene oxide nanocomplex and its application in turn-on detection of metalloproteinase-9. Biosens. Bioelectron. 89: 666–672.

[66] Gu, Y., Q. Wen, Y. Kuang and J. Jiang. 2014. Peptide-templated gold nanoclusters as a novel label-free biosensor for the detection of protease activity. RSC Advances 4: 13753–13756.

[67] Wen, Q., Y. Gu, L. -J. Tang, R. -Q. Yu and J. -H. Jiang. 2013. Peptide-templated gold nanocluster beacon as a sensitive, label-free sensor for protein post-translational modification enzymes. Anal. Chem. 85: 11681–11685.

[68] Wang, H., X Yuan, G Zeng, Y Wu, Y Liu, Q Jiang and S Gu. 2015. Three dimensional graphene based materials: Synthesis and applications from energy storage and conversion to electrochemical sensor and environmental remediation. Adv. Colloid Interface Sci. 221: 41–59.

CHAPTER 9

An Overview of the Effect of Graphene as a Metal Protector Against Microbiologically Influenced Corrosion (MIC)

*Reza Javaherdashti** and *Rahil Sarjahani*

1. Introduction: Microbiologically Influenced Corrosion (MIC)

Corrosion, according to ISO 8044 standard, is defined as "physicochemical interaction (usually electrochemical in nature) between a metal and its environment which results in changes in the properties of the metal and which may often lead to impairment of the function of the metal, the environment, or the technical system of which these form a part" [1]. In the processes of extractive metallurgy, which involves reductive processes, by giving more electrons to metallic compounds in the ore under consideration, thermodynamically stable metal in the ore is brought into a thermodynamically instable state by reducing processes of extractive metallurgy. Consequently, the metals tend to release the additional electrons they received to reach a stable state, and this represents the thermodynamic basis of oxidation or corrosion in general. Microbiologically influenced corrosion (MIC) has been defined in several ways that are more or less similar. Bearing in mind that the term "micro-organism" actually refers to bacteria, cyanobacteria, algae, lichens and fungi, MIC is the term used for the phenomenon in which corrosion is initiated and/or accelerated by the activities of micro-organisms [2].

Microbial corrosion is the corrosion brought about by the activities and presence of microbes. This occurs in several forms and can be managed by traditional control methods and biocides. This process of degeneration chiefly acts on metalloids, metals and rock-based materials. Apart from bacteria, microbial corrosion can also be influenced by micro algae, inorganic and organic chemicals. This form of corrosion

ParsCorrosion, Perth, 6107, WA, Australia.
* Corresponding author: javaherdashti@yahoo.com

affects entities like power plants and chemical industries, as well as facilities that make use of cooling towers.

Microbial corrosion is not caused just by one microbe but can be attributed to several types of microbes. These are usually grouped by their main characteristics like the effect on compounds and by products. In general, the microbes responsible for microbial corrosion can be categorized in two groups according to oxygen requirements:

- Aerobic (needing oxygen): like bacteria that are sulfur oxidizing
- Anaerobic (needing no or little oxygen): like bacteria that are sulfate reducing

Almost all microbial corrosion takes the appearance of pits forming underneath living matter colonies, minerals, and bio-deposits. This results in bio-film formation that results in a confined environment where the conditions can be corrosive. This in turn accelerates the corrosion process.

There are several origins of corrosion, and in turn different types. One of these types of corrosion involves biological processes, where organisms can produce electron flow or modify the local environment to change from a non-corroding to a corrosive one. Some of these processes will be outlined in the following section. When microbial deposits form on the surface of a metal, one case that exists is that they can be regarded as inert deposits on the surface, shielding the area below from electrolyte. A differential aeration cell will form, even in the presence of very small colonies. The area directly under the colony will represent the anode and the metallic surface just outside the contact area will support the reduction of oxygen reaction and will represent the cathode. Metal dissolution will occur under the microbial deposit, similar to pit formation. The density of local dissolution areas should match closely the colony density. Microbial deposits can produce components that will change the local environment and thereby induce corrosion. Both inorganic and organic acids can be produced that will initiate corrosion when they are produced at the colony/metal interface. The production of inorganic acids will also lead to hydrogen ion production which may contribute to hydrogen embrittlement of the metal under the area colonized by microbes.

In anaerobic conditions, some bacteria can reduce the sulfate ion to produce oxygen and the sulfide ions. In iron containing materials, the sulfide ions then combine with ferrous ions to form iron sulfide that leads to metal surface dissolution. The oxygen produced reacts with hydrogen to form water molecules [3]. During iron oxidation, hydrogen ions are produced along with hydroxyl ions by the breakdown of water, and the electrons form atomic hydrogens from hydrogen ions. The reaction is therefore multi-staged depending on anodic, cathodic, water dissociation and bacterial reactions. The source of electrons is the oxidation of the metal, while the electron sink is reduction of hydrogen ions. Some bacteria can directly reduce metal atoms to ions. Impedance spectroscopy is one of the most versatile techniques that are applicable to study bio-corrosion. Potentiodynamic scan experiments are often used to determine the effect of biofilms on both anodic and cathodic behavior. The development of microbial corrosion takes place in three stages:

1. Microbe attachment
2. Growth of initial pit and lump
3. Maturation of lump and pit

Microbial corrosion, also called "bacterial corrosion", "bio-corrosion", "microbiologically influenced corrosion", or "microbially induced corrosion" (MIC), is the corrosion that is caused by microorganisms, usually "chemoautotrophs".[1] It can occur at both metallic and non-metallic materials. Bacteria can be the origin of corrosion: some sulfate-reducing bacteria produce hydrogen sulfide, which can cause sulfide stress cracking. *Acidithiobacillus* bacteria[2] produce sulfuric acid; *Acidothiobacillus thiooxidans* frequently damage sewer pipes. *Ferrobacillus ferrooxidans* directly oxidizes iron to iron oxides and iron hydroxides; the rust particles formed on the RMS Titanic wreck are caused by bacterial activity. Other bacteria produce various acids, both organic and mineral, or ammonia.

In presence of oxygen, aerobic bacteria like *Acidithiobacillus thiooxidans*, *Thiobacillus*[3] *thioparus* and *Thiobacillus concretivorus*, all three widely present in the environment, are the common corrosion-causing factors resulting in biogenic sulfide corrosion.

In absence of oxygen, anaerobic bacteria, especially *Desulfovibrio*[4] and *Desulfotomaculum,* are common. *Desulfovibrio salixigens* requires at least 2.5% concentration of sodium chloride, but *D. vulgaris* and *D. desulfuricans* can grow in both fresh and salt water. *D. africanus* is another common corrosion-causing microorganism. The *Desulfotomaculum* genus comprises sulfate-reducing spore-forming bacteria; *Dtm. orientis* and *Dtm. nigrificans* are involved in corrosion processes. Sulfate-reducers require reducing environment; thus, an electrode potential lower than $-100\,\text{mV}$ is required for them to thrive. However, even a small amount of produced hydrogen sulfide can achieve this shift, so the growth, once

---

[1] An organism that depends on inorganic chemicals for its energy and principally on carbon dioxide for its carbon. Also called *chemolithotroph.* Chemotrophs are organisms that obtain energy by the oxidation of electron donors in their environments.

[2] Acidithiobacillus ferrooxidans (basonym Thiobacillus ferrooxidans) can be isolated from iron-sulfur minerals such as pyritedeposits, oxidising iron and sulfur as energy sources to support autotrophic growth and producing ferric iron and sulfuric acid. Acidithiobacillus thiooxidans (basonym Thiobacillus thiooxidans, Thiobacillus concretivorus) oxidises sulfur and produces sulfuric acid; first isolated from the soil [2], it has also been observed to cause biogenic sulfide corrosion of concrete sewer pipes by altering hydrogen sulfide in sewage gas into sulfuric acid.

[3] A genus of small rod-shaped bacteria that lives in water, sewage, soil, derives energy from oxidation of sulfides, thiosulfates and obtain carbon from carbon dioxide, bicarbonates, and carbonates in solution.

[4] Desulfovibrio (or simply D.) is a genus of Gram-negative sulfate-reducing bacteria. Desulfovibrio species are commonly found in aquatic environments with high levels of organic material, as well as in water-logged soils, and form major community members of extreme oligotrophic habitats such as deep granitic fractured rock aquifers. Like other sulfate-reducing bacteria, Desulfovibrio was long considered to be obligately anaerobic. This is not strictly correct: while growth may be limited, these bacteria can survive in O_2-rich environments. These types of bacteria are known as aero tolerant.

started, tends to accelerate. Layers of anaerobic bacteria can exist in the inner parts of the corrosion deposits, while the outer parts are inhabited by aerobic bacteria. For more details, the readers are encouraged to read a comprehensive book by R. Javaherdashti on MIC [2]. Some bacteria are able to utilize hydrogen formed during cathodic corrosion processes. Bacterial colonies and deposits can form concentration cells, causing and promoting galvanic corrosion [4].

2. Influence of MIC on Industry

Bacterial corrosion may appear in the form of pitting corrosion, for example in pipelines of the oil and gas industry. Anaerobic corrosion is evidenced by the occurrence of layers of metal sulfides and hydrogen sulfide smell. On cast iron, a graphitic corrosion selective leaching may be the result, with iron being consumed by the bacteria, leaving graphite matrix with low mechanical strength in place. Various corrosion inhibitors can be used to combat microbial corrosion. Formulae based on benzalkonium chloride are common in oil field industry. Microbial corrosion can also occur at plastics, concrete and many other materials. Two examples are nylon-eating bacteria and plastic-eating bacteria. Sewer network structures are prone to bio-deterioration of materials due to the action of some microorganisms associated with the sulfur cycle.

Around 9% of damages described in sewer networks can be ascribed to the successive action of microorganisms [2]. Sulfate-reducing bacteria (SRB) can grow in relatively thick layers of sedimentary sludge and sand (typically 1 mm thick) accumulating at the bottom of the pipes and characterized by anoxic conditions. They can grow using oxidized sulfur compounds that are present in the effluent as electron acceptor and excrete hydrogen sulfide (H_2S). This gas is then emitted in the aerial part of the pipe and can impact the structure in two ways: either directly by reacting with the material and leading to a decrease in pH, or indirectly through its use as a nutrient by sulfur-oxidizing bacteria (SOB), growing in toxic conditions, which produce biogenic sulfuric acid. The structure is then submitted to a biogenic sulfuric acid attack. Materials like calcium aluminate cements, PVC or vitrified clay pipe may be substituted for ordinary concrete or steel sewers that are not resistant in these environments.

Many industries have been affected by MIC including:

- Chemical processing industries: stainless steel tanks, pipelines and flanged joints, particularly in welded areas after hydrotesting with Natural River or well waters.
- Nuclear power generation: carbon and stainless-steel piping and tanks, copper-nickel, particularly during hydro-test and outage periods.
- Onshore and offshore oil and gas industries: mothballed and water flood systems, oil and gas handling systems, particularly in those environments soured by sulfate reducing bacteria (SRB)—produced sulfides.
- Underground pipelines industry: water—saturated clay-type soils of near—natural pH with decaying organic matter and a source of SRB.

- Water treatment industry: heat exchangers and piping.
- Sewage handling and treatment industry: concrete and reinforced concrete structures.
- Highway maintenance industry: culvert piping.
- Aviation industry: aluminum integral wing and fuel storage tanks.
- Metal working industry: increased wear from breakdown of machining oils and emulsions.
- Marine and shipping industry: acceleration damage to ship and barges.

Positive identification of microbiologically influenced corrosion requires chemical, biological and metallurgical analysis of the waters, soils and metal samples. Microbial corrosion can be a severe problem in inactive water systems. Utilizing mechanical cleaning techniques and biocides can lessen microbial corrosion. However, any area collecting stagnant water is very susceptible to microbial corrosion. Furthermore, microorganisms, that can utilize hydrocarbons, like pseudomonas aeruginosa,[5] can be found in aviation fuel. These microorganisms form dark brown or green mats similar to a gel, and lead to microbial corrosion on the rubber and plastic parts of the fuel system of an aircraft.

The huge economic impact of the corrosion of metallic structures is a very important issue for all modern societies. Estimates for the cost of corrosion degradation run to about €200 billion a year in Europe and over \$270 billion a year in the U.S. The annual cost of corrosion consists of both direct costs and indirect costs. Worldwide, it is estimated that these costs approach \$1 trillion annually. The direct costs are related to the costs of design, manufacturing, and construction in order to provide corrosion protection, while the indirect costs are concerned with corrosion-related inspection, maintenance and repairs [5].

3. Strategies for Prevention of MIC

One successful and generic way to prevent corrosion, no matter what kind of corrosion is, is to cover the open surface of material, here refers to metals, by protective barrier (layer or layers) and limit access of destroying agents. Corrosion processes develop fast after disruption of the protective barrier and are accompanied by a number of reactions that change the composition and properties of both the metal surface and the local environment, such as formation of oxides, and diffusion of metal cations into the coating matrix, local pH changes, and electrochemical potential [6]. The study of corrosion of mild steel and iron is a matter of tremendous theoretical and practical concern and as such has received a considerable amount of interest. This may require the use of corrosion inhibitors in order to restrain corrosion attack on metallic materials.

---

[5] Pseudomonas aeruginosa is a common Gram-negative, rod-shaped bacterium that can cause disease in plants and animals, including humans. It is citrate, catalase, and oxidase positive. It is found in soil, water, skin flora, and most man-made environments throughout the world. It thrives not only in normal atmospheres, but also in low-oxygen atmospheres, and has thus colonized many natural and artificial environments.

Over the years, considerable efforts have been deployed to find suitable corrosion inhibitors of organic origin in various corrosive media. In acidic media, nitrogen-base materials and their derivatives, sulfur-containing compounds, aldehydes, thioaldehydes, acetylenic compounds, and various alkaloids (for example, papaverine, strychnine, quinine and nicotine) are used as inhibitors [7]. In neutral media, benzoate, nitrite, chromate, and phosphate act as good inhibitors. Inhibitors decrease or prevent the reaction of the metal with the media. They may slow down or reduce the corrosion rate by performing the following effects:

- Adsorbing ions/molecules on to metal surface
- Increase or decrease the reaction (anodic and/or cathodic)
- Decrease diffusion rate
- Decrease the electrical resistance of the metal surface
- Inhibitors that are often easy to apply and have *in situ* application advantage

Several factors including cost, amount, availability and most importantly safety to environment need to be considered when choosing a barrier layer. It is good to mention here the possible roles of inhibitors:

- Adsorption of the molecules (or ions) on active sites
- Increase in cathodic and/or anodic overvoltage
- Formation of a protective barrier film

Some important factors can contribute to the action and influence of inhibitors:

- Chain length
- Size of the molecule
- Bonding nature
- Strength of bonding to the substrate
- Cross-linking ability
- Solubility in the environment

The role of inhibitors is to form a barrier of one or several molecular layers that can act as a shield against acid attack. This protective action is often associated with chemical and/or physical adsorption, involving a variation in the charge of the adsorbed substance and transfer of charge from one phase to the other. Sulfur and/or nitrogen-containing heterocyclic compounds with various substituents are considered to be effective corrosion inhibitors. For instance, thiophene and hydrazine derivatives offer special affinity to inhibit corrosion of metals in acidic solutions. Inorganic substances such as phosphates, chromates, dichromates, silicates, borates, tungstates, molybdates and arsenates have been found effective as inhibitors of metal corrosion. Pyrrole and its derivatives are believed to exhibit good protection against corrosion in acidic media [6].

These inhibitors have also found useful application in the formulation of primers and anticorrosive coatings, but a major disadvantage is their toxicity and as such their use has come under severe criticism. Among the alternative corrosion inhibitors, organic substances containing polar functions with nitrogen, sulfur, and/or oxygen in the conjugated system have been reported to exhibit good inhibiting properties. The inhibitive characteristics of such compounds are originated from the adsorption ability of these molecules, with the polar group acting as the reaction center for the adsorption process. The resulting adsorbed film acts as a barrier that separates the metal from the corrodent, and thus the efficiency of inhibition depends on the mechanical, structural and chemical characteristics of the adsorption layers formed under particular conditions.

Inhibitors are often added in industrial processes to secure metal dissolution in acid solutions. Standard anti-corrosion coatings which are developed to date passively prevent the interaction of corrosive species with the metal. The known hazardous effects of most synthetic organic inhibitors and the need to develop cheap, nontoxic and eco-friendly processes have now urged researchers to focus on the use of natural products as inhibitors. Currently, there is an urgent need to develop sophisticated new generation coatings for improved performance, especially in view of Cr (VI)-based inhibitor being banned and labeled as a carcinogen. The use of inhibitors is one of the best options of protecting metals against corrosion. Several inhibitors in use are either synthesized from cheap raw materials or chosen from compounds having heteroatoms in their aromatic or long-chain carbon system. However, most of these inhibitors are toxic to the environment. This has prompted the search for green corrosion inhibitors. Among them, graphene has recently been considered as an effective alternative coating for corrosion inhibition.

4. Graphene: A Promising Corrosion Inhibitor Candidate

New research features graphene as a promising novel surface coating that can be used to minimize metallic corrosion, especially under harsh microbial conditions [8]. The most significant finding is that graphene coating offers 100-fold improvement in corrosion resistance compared to commercial polymer coatings which are available in the market. This finding is remarkable considering that graphene is nearly 4000 times thinner than several commercial coatings but offers more than an order of magnitude higher resistance to microbial attack [8]. Graphene is an allotrope (form) of carbon consisting of a single layer of carbon atoms arranged in a hexagonal lattice, as shown in Fig. 1. It is the basic structural element of many other allotropes of carbon, such as graphite, charcoal, carbon nanotubes and fullerenes. It can be considered as an indefinitely large aromatic molecule, the ultimate case of the family of flat polycyclic aromatic hydrocarbons.

In more complex terms, it is an allotrope of carbon in the structure of a plane of sp^2 bonded atoms with a molecule bond length of 0.142 nm. Layers of graphene stacked on top of each other form graphite, with an interplanar spacing of 0.335 nm (see Fig. 1)

Graphene is the thinnest compound known to man at one atom thick, the lightest material known (with 1 square meter coming in at around 0.77 mg), the strongest compound discovered (between 100–300 times stronger than steel and with a tensile stiffness of 150,000,000 psi), the best conductor of heat at room temperature (with a thermal conductivity of $(4.84 \pm 0.44) \times 10^3$ to $(5.30 \pm 0.48) \times 10^3$ W m^{-1} K^{-1}) and also the best conductor of electricity known (studies have shown electron mobility

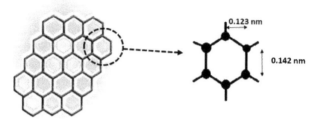

Fig. 1. Schematic of a graphene layer and a single hexagonal lattice of carbon atoms.

at values of more than 15,000 cm^2 V^{-1} s^{-1}). Other notable properties of graphene are its unique levels of light absorption at $\pi\alpha \approx 2.3\%$ of white light and its potential suitability for use in spin transport [9, 10].

Bearing this in mind, you might be surprised to know that carbon is the second most abundant mass within the human body and the fourth most abundant element in the universe (by mass), after hydrogen, helium and oxygen. This makes carbon the chemical basis for all known lives on earth, therefore graphene could be an ecologically friendly, sustainable solution for an almost limitless number of applications. Since the discovery (or more accurately, the mechanical fabrication) of graphene, advances within different scientific disciplines have been explored, with huge gains being made particularly in electronics and biotechnology fields. The electronic structure of graphene is rather different from usual three-dimensional materials. A. Geim, Noble Laureate in 2010 for his work on graphene, and his co-workers had reported in 2009 the electronic properties of graphene [10]. Graphene's Fermi surface is characterized by six double cones, as shown in Fig. 2. In intrinsic (undoped) graphene, the Fermi level is situated at the connection points of these cones. Since the density of states of the material is zero at that point, the electrical conductivity of intrinsic graphene is quite low and is in the order of the conductance quantum ($\sim 2e^2/h$, where e is the elementary charge and h is the Plank constant); the exact pre-factor is still debated. The Fermi level can however be changed by an electric field so that the material becomes either n-doped (with electrons) or p-doped (with holes) depending on the polarity of the applied field. Graphene can also be doped by adsorbing, for example, water or ammonia on its surface. The electrical conductivity for doped graphene is potentially quite high; at room temperature, it may even be higher than that of copper. Close to the Fermi level, the dispersion relation for electrons and holes is linear. Since the effective masses are given by the curvature of the energy bands, this corresponds to zero effective mass. The equation describing the excitations in graphene is formally identical to the Dirac equation for massless fermions which travel at a constant speed. The connection points of the cones are therefore called Dirac points. This gives rise to interesting analogies between graphene and particle physics, which are valid for energies up to approximately 1 eV, where the dispersion relation starts to be nonlinear.

The unit hexagonal cell of graphene contains two carbon atoms and has an area of 0.052 nm^2. We can thus calculate its density as being 0.77 mg/m^2. A hypothetical hammock measuring 1 m^2 made from graphene would thus weigh 0.77 mg. Graphene

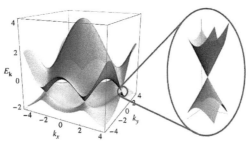

Fig. 2. Schematic bandgap of graphene. Reprinted with permission from Ref. [10].

is almost optically transparent [10]. This number is given by $\pi\alpha$, where α is the fine structure constant. Thus, suspended graphene does not have any color. Graphene has a breaking strength of 42 N/m. Steel has a breaking strength in the range of 250–1200 MPa, which corresponds to $0.25–1.2 \times 10^9$ N/m². For a hypothetical steel film of the same thickness as graphene (which can be taken to be 3.35 Å = 3.35×10^{-10} m, i.e., the layer thickness in graphite), this would give a 2D breaking strength of 0.084–0.40 N/m. Thus, graphene is more than 100 times stronger than the strongest steel. In our 1 m² hammock tied between two trees, you could place a weight of approximately 4 kg before it would break. It should thus be possible to make an almost invisible hammock out of graphene that could hold a cat without breaking. The hammock would weigh less than one mg, corresponding to the weight of one of the cat's whiskers.

The thermal conductivity of graphene is dominated by phonons and has been measured to be approximately 5000 W m⁻¹ K⁻¹. Copper at room temperature has a thermal conductivity of 401 W m⁻¹ K⁻¹. Thus, graphene conducts heat 10 times better than copper. As summary, some of graphene properties are listed below [11]:

- *Mechanical strength*: 200 times stronger than steel
- *Thickness*: 1 million times thinner than a hair
- *Conductivity*: the world's most conductive material
- *2-dimensional structure*
- *Stretchable, flexible*
- *Promising*: more than 200 universities in the world are working on it
- *Ultra-light*
- *Transparent*
- *Waterproof*

5. Recent Advances on the Application of Graphene as a Smart Anti-corrosive Coating for MIC

Graphene has recently showed superiority over some polymer coatings for prevention of MIC. Krishnamurthy et al. have investigated the effectiveness of nano-scale graphene coatings in prevention of MIC of a metallic structure under galvanic conditions. More specifically, they compared the MIC resistance of Ni in the presence of three different coatings, namely parylene-C coating on Ni (PA/Ni), polyurethane coating on Ni (PU/Ni) and graphene coating on Ni (Gr/Ni) [8]. The study showed that graphene coatings outperform its polymer counterparts (parylene-C and polyurethane) by offering 1 and 2 orders of magnitude better impedance against microbial corrosion. The shortcomings associated with defect-based corrosion of graphene coatings in abiotic environments have been countered in their study by the use of 3–4 layers of graphene films that are grown *in situ* with minimal defect density.

Further, the few defects that may be present on the surface can be plugged by biofilms through the production of non-conductive polymeric debris. On exposure to similar harsh microbial environments, parylene-C failed due to its extensive microbial degradation and polyurethane failed due to its non-conformal adhesion to the metallic surface. In order to study the feasibility of graphene coatings as MIC resisting element on non-CVD (chemical vapor deposition) surfaces, they also studied the effect of wet transfer on the quality of graphene. The study revealed that transferred graphene films are significantly more defective than their as-grown counterparts, thereby future research should be targeting defect-free transfer of graphene for surfaces that are not amenable to CVD growth.

When metals are in contact with biofilms, which are naturally formed under the exposure of materials to ambient moisture, miscellaneous metabolic activities lead to the production of organic and inorganic acids, volatile compounds and other chemical reactions onto the metals. This phenomenon represents an enormous burden to maintenance cost in metal-based infrastructure. There are ways to prevent microbial corrosion in metallic structures via attacking the biofilms that cause microbial corrosion. They mainly include physical methods (e.g., flushing) and chemical methods (e.g., biocides).

The use of protective epoxy coatings in order to isolate metals from environment and the application of passivation layers (e.g., thiol-based mono-layers) have also been considered as standard mitigation solutions [11]. However, compared to the lifetime of the objects being protected, all these control approaches exhibit a short-term efficiency, deteriorating and roughening as they age. In addition, biocides and organic solvents present in protective layers have a negative impact in complete ecosystems where biofilms are treated. With the pressure of stringent environmental regulations monotonically increasing, there is an urgent need for new environmentally friendly and sustainable microbial corrosion control strategies. Moreover, there is a phenomenon, similar to microbial corrosion, that includes metal ion release when a metallic material is in contact with other living systems, such as the one provided by human skin.

Among metallic materials, Ni and its alloys are one of the most widely used in diverse domestic objects that come in contact with human skin. Ni has been chosen for most of these applications due to its mechanical properties, corrosion resistance, durability, appearance and availability. However, remarkable as these properties are, they do not seem to withstand direct interaction with human skin conditions. Release of Ni^{2+} ions from Ni metallic objects that come in direct contact with human skin has been related to a phenomenon known as the allergic contact hypersensitivity response to Ni. This is one of the most frequent contact allergies produced by man-made products and a major source of this Ni hypersensitivity [11].

Different nano-scale approaches have been used to produce certain degrees of control over contact microbial corrosion issues. Although such approaches seem to be effective, they raise concerns about the risks of absorption of nanoparticles through the skin. Similarly, metal nanoparticle-based coatings have been also used for microbial corrosion control. However, these coatings fall into the biocide category, exhibiting a strong bactericide effect coming from the release of metallic ions from the nanoparticles. Recently, it has been reported that Cu microbial corrosion passivation using single-layer CVD graphene (SLG) grown on that metal was efficient [11]. Although SLG effectively protects copper, coupling between graphene and copper electronic states is very weak compared to electronic interfacial bonding in other graphene-metal systems, such as graphene on Ni or graphene on Ti/Au. Electronic coupling becomes especially relevant when biological response of CVD graphene-coated metals is considered due to the reported connection between electron transfer from these materials to microorganisms and cell death.

When protection of metals (other than Cu) from biological interaction using graphene coatings is pursued, a different and an interesting scenario is found. It is important to highlight that microbial corrosion cannot be completely linked to a single chemical reaction because it is influenced by the complex processes of microorganisms performing different electrochemical reactions and secreting proteins and metabolites that have effects over a material's damage.

Although corrosion-resisting properties of graphene coatings have been widely studied, evaluation of graphene performance under microbiologically-assisted corrosion conditions presents a different scenario, especially considering that graphene/metal system performance depends on graphene and the underlying metal electronic interaction. From the biological point of view, graphene grown on Ni, Cu or any other metal or alloy behaves differently. In the case of Ni, for instance, graphene's electronic structure is disturbed through a strong hybridization of the metal's d-orbital and graphene's orbital forming a chemisorption interface. In this scenario, the strong coupling between Ni and graphene electronic states might play a major role in the biological response of the nanoscale-modified materials that need to be addressed [11].

Motivated by the need of studies exploring the protection that graphene coatings offer to other relevant metals (other than copper), a study on the efficiency of graphene grown and transferred on Ni as an ionic barrier under two biological conditions, microbial corrosion and sweat immersion, has been conducted [11]. Chemical vapor deposition (CVD) graphene, with surface areas in the centimeter square range, has been chosen as a coating due to its versatility and lack of intrinsic cytotoxicity. This condition is opposite to its close relative, graphene oxide (GO), formed by micro- or nano-sized flakes of functionalized graphene in powder or solution, which has been shown to present an antibacterial activity. Parra et al. [11] explored the efficiency of as-grown few layer graphene (FLG) and transferred graphene onto Ni foils. They mainly focused on the way in which these nanostructured coatings modify the interaction between: (1) Ni and bacteria (to test microbial corrosion passivation of such coatings) and (2) Ni and sweat (to evaluate Ni ion release from coated samples in conditions given by standard tests to quantify material biocompatibility). On the other hand, studies on rGO[6] multi layers have been conducted by Kang et al. [12] using layer-by-layer assembly that showed acceptable resistance to corrosion. Recent studies have also reported the development of anti-corrosive coatings based on graphene and graphene-polymer composite [13, 14].

The oxidation resistance of Fe and Cu foils by the rGO layers was confirmed by Raman spectroscopy, optical microscopy and SEM [12]. The results revealed that the rGO layers play a role as a diffusion barrier of gas. The solution process using rGO is easy and reproducible. Forming a gas proof layer on the metal surface can always help in prevention of any kind of chemical attack help in the decay of the substrate.

Among all the properties of graphene, the one that seems to have more relevance to industry is its stability. More specifically, graphene has been shown to be a good anti-corrosion and anti-oxidation material. The use of graphene in this field is one of the easiest possible approaches of coatings, since it doesn't need any additional treatment, device pattern or lithography; the raw material should just be grown on or transferred onto the metal to protect. This industry-friendly application may be the motor field for the introduction of graphene in mass consumables.

From the point of view of impermeability, graphene can be understood as a nanometric shield which can effectively withstand the wear produced by invasive agents from ambient conditions. The hexagonal network of carbon atoms in graphene is so dense that no other known material can penetrate it easily. Bunch and co-workers [15] described graphene as a membrane that could block transferring gas molecules. The graphene film could withstand a pressure difference larger than one atmosphere. Figure 3a schematically shows a graphene-sealed SiO_2 micro-chamber.

Graphene thickness was altered from 1 to $\sim$ 75 layers, which was proved by Raman spectra. The graphene film was able to stick to the micro-chamber side walls by van der Waals forces between graphene and SiO_2, forming a gas package with a size of $\sim$ (μm^3). The inset in Fig. 3a shows a 4.75 μm $\times$ 4.75 μm squared graphene sheet sealing the top face of the micro-chamber. The pressure difference Δp was defined as the difference value between the inside pressure (P_{int}) (atmospheric pressure, 101 kPa) and the external pressure (P_{ext}) of the micro chamber, $\Delta p = P_{int} - P_{ext}$. If P_{ext} is changed, P_{int} would accordingly change to a balance state with P_{ext} because of the relaxation effect, and the relaxation time could be varied from minutes to days depending on the gas species and the temperature.

For times shorter than the equilibration time, an apparent pressure difference Δp can exist across the membrane, causing it to stretch like the surface of a balloon (Fig. 3b). Situations for $\Delta p > 0$ and $\Delta p < 0$ are shown in Figs. 3c and 3d, respectively. The authors successfully created a positive pressure difference ($\Delta p > 0$) and a lower pressure difference in the chamber ($\Delta p < 0$) and observed images of the membrane bulged upward and deflected downward by tapping mode atomic force microscope (AFM), respectively. Figure 3e shows results for different membranes of altered thicknesses and for different gases: air, argon and helium. Each gas showed no

---

[6] Reduced graphene oxide.

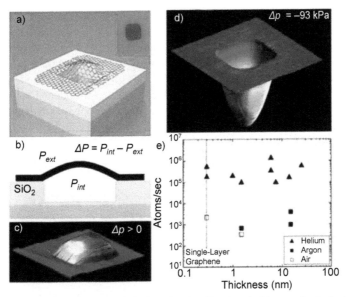

Fig. 3. (a) Schematic of a graphene sealed micro chamber. (Inset) optical image of a single atomic layer graphene drumhead on SiO_2. (b) Side view schematic of the graphene sealed micro-chamber. (c) AFM image of a ~ 9 nm thick many-layer graphene drumhead with $\Delta p > 0$. (d) AFM image of the graphene sealed micro-chamber of Fig. 3a with $\Delta p = -93$ kPa across it. (e) Relationship of the gas leak rates vs. thickness: Helium (▲), argon (■) and air (□). Reprinted with permission from Ref. [15].

noticeable dependence on thickness from 1 to 75 atomic layers under the situation of approximately same pressure difference applied across the membrane, indicating that the leak was neither through the graphene sheets nor through defects in these sheets. This suggests that glass walls of the micro-chamber or the graphene-SiO_2 sealed interface should be responsible for the gas leak. This offers the opportunity to probe the permeability of gases through atomic vacancies in single layers of atoms and defects patterned in the graphene membrane, acting as selective barriers for ultrafiltration.

Graphene has also shown the ability of protecting some materials from oxidation. Chen et al. [16] used as-grown graphene in CVD system for the first time to protect Cu or Cu/Ni alloy surfaces from oxidation. The introduction of this protection layer led to a passivation of the metal or metal alloy underneath. It is proved that graphene transferred from original growth substrate to the target substrate also has a passivation effect. The above discussed impermeability is one of the aspects responsible for the passivation since the covering graphene offered an inserted block between the reactants and the protected metal or metal alloy. In other words, a solid film barrier was provided to separate the environment and the metal surfaces physically. On the other hand, the employed barrier layer should be chemically inert in the active oxidation environments. Therefore, the stability of graphene in wild oxidation environments is another important consideration for its further applications. To investigate and verify graphene's chemical stability, graphene-covered Cu and Cu/Ni samples were exposed to accelerated oxidative environments, one of which was lengthy air annealing (200°C, 4 h) and the other one was hydrogen peroxide environment (H₂O₂, 30%, 2 min). Figure 4a schematically shows that the graphene film played the role of a molecular diffusion barrier, keeping the reactive agent away from the metal underneath. Interestingly, as a vivid example, comparison was carried out between two penny coins (95%Cu /5%Zn), one of which was coated by graphene monolayer and the other was naked. H₂O₂ treatment was introduced to these coated and uncoated coins. As shown in Fig. 4b, the uncoated copper penny showed a color change caused by oxidation, whereas the graphene-coated coin kept the appearance like its original.

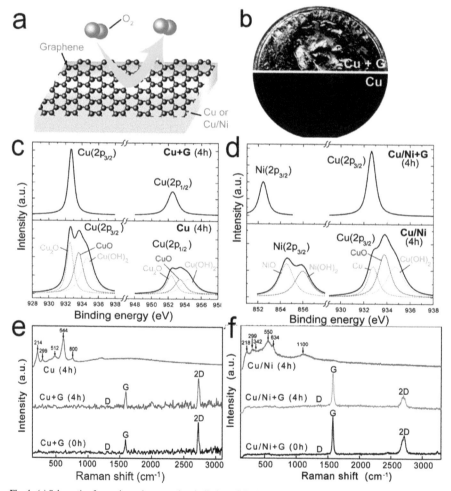

Fig. 4. (a) Schematic of a graphene sheet as a chemically inert diffusion barrier. (b) Image of a penny coated (upper) and uncoated (lower) with graphene after treated in H₂O₂ for 2 min (30%). (c) XPS core-level Cu₂p spectrum of coated (upper) and uncoated (lower) Cu foil after air anneal (200°C, 4 h). (d) XPS core-level Ni 2p₃/₂ and Cu 2p₃/₂ spectrum of coated (upper) and uncoated (lower) Cu/Ni foil after air anneal (200°C, 4 h). Raman spectrum of the (e) Cu and (f) Cu/Ni alloy foils with and without a graphene coating after air anneal (200°C, 0 and 4 h, respectively). Reprinted with permission from Ref. [16] Copyright 2011 American Chemical Society.

Graphene offers an impenetrable barrier for external agents but, surprisingly, it also presents a more exotic property: wetting transparency. The chemical interaction between a graphene-coated material and environment is unignorable. X-Ray Photoelectron Spectroscopy (XPS) analysis was able to monitor elemental variations before and after heating (200°C, 4 h) in both graphene-coated and uncoated Cu and Cu/Ni substrates. Before air annealing, original graphene-coated and uncoated Cu possessed same sharp peaks in their XPS spectra, at binding energies of 932.6 and 952.5 eV, which has been proved corresponding to $Cu(2p_{3/2})$ and $Cu(2p_{1/2})$, respectively.

Condensation heat transfer experiments on a Cu sheet (40 mm in diameter) with and without monolayer-graphene deposition were carried out to test the function of graphene. Figures 4c and d show changes in the electronic states of bare Cu after 4h of annealing, corresponding to oxides formation. On the other hand, the graphene-coated Cu samples did not show such changes, suggesting the efficient role of graphene as a barrier. Surface oxides and strong variation of surface wettability may introduce negative impacts because oxides play the role of a thermal barrier that reduces the heat transfer coefficient at Cu interfaces. This phenomenon can turn the material surface into hydrophilic, which can lead to a liquid film which adversely affects condensation heat transfer. The condensation heat transfer was enhanced by approximately 30–40% (over a wide range of temperatures) by coating Cu with graphene.

Understanding graphene's wettability opens a new field for its applications. Specially, when oxygen-containing functional groups are introduced in the basal plane, graphene can be tuned hydrophilic. Along with increasing the oxygen-containing functional group concentration, the water contact angle decreases. On the other hand, hydrophobic effect can be achieved by electrical doping of single layer graphene, which contributes to renormalization of the surface tension between graphene and water. In the same line, artificial super hydrophobic graphene was also reported. By due to the wettability of graphene, not only wetting transparency can be carried out, but also slip flow inside the interlayer gallery between graphene layers can be optionally enhanced and broken down. Graphene single layer was reported to be successfully used as a coating material to prevent oxidation of different underlying substrates [16]. As mentioned, the genuine sp^2 hexagonal distribution of carbon atoms provides an impermeable physical barrier that avoids substrate-environment interaction, as already demonstrated in other ultra-thin carbon-based structures such as carbon nanotubes. Based on their results, Chen et al. have demonstrated that graphene can provide effective oxidation resistance for the underlying Cu and Cu/Ni alloys [16].

It has been demonstrated that the oxidation stability of graphene is strongly related to its nanostructure. Figure 5 shows an optical micrograph of the graphene film transferred onto a SiO_2/Si substrate from the original deposited Cu foil. Graphene flakes synthesized by any method always contain a certain degree of non-ideality, such as domain boundaries rich in pentagonal and heptagonal lattices, missing bonds, impurities and wrinkles, leading to an inhomogeneous degree of protection. The first sign of non-ideality of graphene coating dates to 2009 [17]. In the same

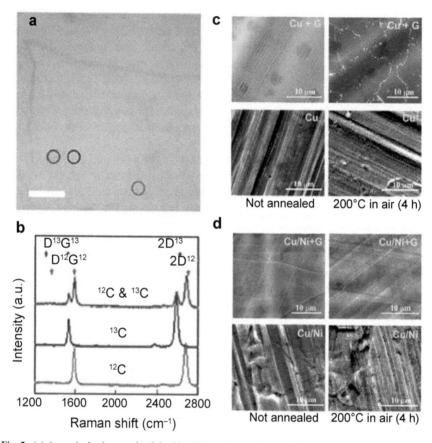

Fig. 5. (a) An optical micrograph of the identical region analyzed with micro-Raman spectroscopy. (b) Raman spectra from ^{12}C-graphene (green circle), ^{13}C-graphene (blue circle), and the junction of ^{12}C- and ^{13}C-graphene (red circle), respectively. (c) SEM images of graphene coated (upper) and uncoated (lower) Cu foil taken before (left column) and after (right column) annealing in air. (d) SEM images of graphene coated (upper) and uncoated (lower) Cu/Ni foil taken before (left column) and after (right column) annealing in air. Reprinted with permission from Ref. [16, 18]. Copyright 2009 American Chemical Society and 2011 American Chemical Society.

work in which the anti-oxidation property of graphene was reported [16], the authors suggested that the local oxidation of the protected material can take place at the graphene defects, mainly located at the domain boundaries. More specifically, the SEM micrograph had a number of small bright white spots representing the formed oxides (Fig. 5), most likely at the graphene grain boundaries or defect sites of the graphene surface. Similarly, short time exposure to the oxidizing aqueous solution of H_2O_2 also showed significant attack in few spots (white regions) after 15 and 5 min of H_2O_2 exposure, respectively.

The intrinsic inhomogeneity of graphene leads to a diversity of surface defects that act differently in contact with the environment (gas and liquids). Therefore, the introduction of graphene in real applications requires an exhaustive characterization of these features, including reliability models and lifetime predictions. It seems evident that the changes produced on graphene-coated metals are related to local oxidation.

The stress occurred at defect boundaries in graphene can be sorted as zigzag tilt boundaries—armchair tilt boundaries—domains, and the energy for each situation is ranked in the same manner. Thus, the zigzag tilt boundaries are the most sensitive to the environment variations, followed by the armchair tilt boundaries. The combination of zigzag tilt boundaries and armchair tilt boundaries causes the formation of defect boundaries of graphene. This also explains why the oxidation preferentially happens at the corners of the domain boundaries than along with the boundaries. The results are consistent with the observations mentioned in Fig. 5.

6. Graphene as a Coating: Challenges and Approaches to Commercialization

As discussed above, graphene is also oxidized, which might limit its applications. In fact, all carbon materials, including carbon nanotubes and graphene, are not stable as their ends or edges are readily oxidized owing to the presence of more defects, such as dangling bonds in regions that are susceptible to be attacked by OH⁻ ions. In the field of using graphene as a protective coating, the key factor defining its performance is the sheet quality. The main factor that reduces the performance of graphene is its amount of defects. The most common non-idealities in graphene sheets are: (i) missing bonds, (ii) presence of pentagonal and hexagonal lattices, (iii) lattice distortions, (iv) local thickness fluctuations, and (v) doping with impurities. All these factors can not only alter the properties of graphene, but can also represent centers for damage accumulation when using it as a protective coating. Local defects in graphene can lead to accumulation of oxygen, which impoverishes the mechanical, electronic and chemical properties of the sheet. On the contrary, the pristine hexagonal structure of graphene forms a perfect impermeable barrier for material protection. Undoubtedly, the material still needs to be ameliorated in the sense of defects confinement, and therefore, the future improvement of graphene coatings seems to be linked to the reduction of defects. One of the main targets is to minimize the amount of graphene domain boundaries, since they contain most of the defects. Moreover, using selective substrates that minimize the introduction of impurities in the carbon lattice, to grow the material on it by CVD method is also highly desirable. Finally, the production of large area sheets is one of the main requirements for its use as a protective coating, since the areas to be protected are normally at the macroscopic scale.

Led by Lingxue Kong, Australian scientists have demonstrated that graphene flakes grown directly onto stainless steel makes it resistant to corrosion, drawing the attention of the coatings industry [12]. In their work, this team at Deakin University, Australia grew 3D networks of graphene nanoflakes on micron-sized fibers of stainless steel, improving both the metal's resistance to corrosion and its electrical conductivity. Corrosion is a complex process that is linked to both environmental factors and the conditions of the metal surface (e.g., roughness, surface area and presence of oxides). Although stainless steel is more resistant to corrosion than other metals in acidic aqueous environments, it can be susceptible to localized forms of attack that can result in cracking. Deakin's work [12] utilizes the natural hydrophobic properties of graphene to repel water from the surface of stainless-steel fibers. In addition, graphene's electrical properties reduce the likelihood of redox reactions occurring on the surface, preventing oxidation of the substrate metal. The team coated highly porous austenitic substrates—consisting of evenly distributed 15 μm diameter steel fibers—with networks of carbon nanoflakes. By tuning the temperature and feed gas flow rate during growth, the thickness of graphene coating could be tightly controlled. The resulting coating varied from a few atomic layers thick, to complex, interconnected nano-pillars of carbon which increased the specific surface area of the material by up to 26,000 times.

The inhomogeneity of stainless steel's microstructure was found to produce a graphene coating that was not highly crystalline. In addition, the wettability of the material was also studied, and it was found that when the density of carbon nano-pillars was the highest, the coated steel was super-hydrophobic. The material also displayed its highest corrosion resistance to synthetic seawater at this point, suggesting that it was the presence of graphene that increased its corrosion resistance, without compromising the properties or structure of the native stainless-steel material. The team expects potential applications of their coated-steel to include thermal exchangers, molecular separation systems and bio-compatible materials [19].

In this chapter, we highlighted the characteristics and novel applications of graphene-based materials, focusing on its use as a protective coating. Graphene as an anti-corrosion coating is a very attractive candidate because it may protect many metals by keeping their intrinsic properties unaltered, which is something that cannot be achieved using three dimensional protective paints, oxides or polymers. Moreover, the use of graphene in this field doesn't require device patterning, which could decrease the quality of the two-dimensional sheet. However, the main problem is that the graphene sheets synthesized using the current methods still contain too many defects, which can lead to an imperfect protection, especially in long-term experiments. Therefore, the aging of graphene and the reliability tests of graphene-based prototypes is an essential step prior to their introduction in the industry, and many engineers and scientists will have to face it in the next years.

Hardly a day goes by without mention of another potential application for graphene. We have addressed in this chapter its application as anti-corrosive coating. Its unique combination of properties has promoted its use in batteries, electronics, filtration systems, composites and coatings.

7. Conclusions

Corrosion and particularly microbiologically influenced corrosion (MIC) have engineering importance as assessed by both their high costs and risks involved. In other words, the significance of MIC is not only limited to its economic-environmental costs but also to the risks that it can create (or, in the case of already existing risks, enhance) such as, but not limited to, contamination, accelerated corrosion, increasing degradation, decreasing reliability and even causing safety threats such as those to be encountered in systems such as fire water rings.

Being electrochemical in nature, MIC can be dealt with in the following five strategies:

1) Application of chemical treatment (such as biocides)
2) Application of cathodic protection
3) Mechanical-physical mean and/or barriers such as pigging or coating
4) Design factor to control the MIC risk
5) Application of biological means (biological treatment of MIC)

In this chapter, we only discussed the option of creating a physical barrier between the corrosive environment (as enhanced with the existence and activity of corrosion-related bacteria and corrosion-related archaea) and the metal surface. We focused here on the performance of graphene as the coating.

Currently, the coatings that are used to control MIC are polymeric materials such as epoxies or polyurethane. These coatings could themselves become vulnerable to degradation by the bacteria over time, as has recently been evidenced in the case of epoxy. However, graphene seems to be a suitable coating material that, in the light of its properties and what has been researched so far, possesses properties good enough to be considered as the future coating material that would be quite resistant to biodegradation.

In this chapter, after we briefly discussed corrosion and MIC, we addressed some important properties of graphene which endow graphene as a potential coating material and an alternative to what is currently used in industrial applications. High resistance towards MIC could be a promising feature of graphene that would make it an ideal coating material for a series of industries ranging from oil and gas to marine industry and water/waste-water treatment plants.

References

[1] Corrosion of metals and alloys—Basic terms and definitions, ISO 8040:2015 (E/F), ISO, Switzerland, 2015.
[2] Javaherdashti, R. 2008. Microbiologically Influenced Corrosion: An Engineering Insight. London.
[3] Gujarathi, K. 2008. Corrosion of Aluminum Alloy 2024 Belonging to the 1930s in Seawater Environment. Texas A&M University, pp. 98.
[4] Schwermer, C. U., G. Lavik, R. M. M. Abed, B. Dunsmore, T. G. Ferdelman, P. Stoodley, A. Gieseke and D. de Beer. 2008. Impact of nitrate on the structure and function of bacterial biofilm communities in pipelines used for injection of seawater into oil fields. Applied and Environmental Microbiology 74: 2841–2851.
[5] Berger, M. 2008. Self-healing nanotechnology anticorrosion coatings as alternative to toxic chromium. Nanowerk. https://www.nanowerk.com/spotlight/spotid=6555.php, accesssed 29.04.2019.
[6] Adejoro, I. A., F. K. Ojo and S. K. Obafemi. 2015. Corrosion inhibition potentials of ampicillin for mild steel in hydrochloric acid solution. Journal of Taibah University for Science 9: 196–202.
[7] Rani, B. E. A. and B. B. J. Basu. 2012. Green inhibitors for corrosion protection of metals and alloys: an overview. International Journal of Corrosion 2012: 380217.
[8] Krishnamurthy, A., V. Gadhamshetty, R. Mukherjee, B. Natarajan, O. Eksik, S. Ali Shojaee, D. A. Lucca, W. Ren, H. -M. Cheng and N. Koratkar. 2015. Superiority of graphene over polymer coatings for prevention of microbially induced corrosion. Scientific Reports 5: 13858.
[9] Scientific Background on the Nobel Prize in Physics 2010: Graphene, Royal Swedish Academy of Sciences, 2010.

[10] Castro Neto, A. H., F. Guinea, N. M. R. Peres, K. S. Novoselov and A. K. Geim. 2009. The electronic properties of graphene. Reviews of Modern Physics 81: 109–162.

[11] Parra, C., F. Montero-Silva, D. Gentil, V. Del Campo, T. Henrique Rodrigues da Cunha, R. Henríquez, P. Häberle, C. Garín, C. Ramírez, R. Fuentes, M. Flores and M. Seeger. 2017. The many faces of graphene as protection barrier. performance under microbial corrosion and Ni allergy conditions. Materials 10: 1406.

[12] Kang, D., J. Y. Kwon, H. Cho, J. -H. Sim, H. S. Hwang, C. S. Kim, Y. J. Kim, R. S. Ruoff and H. S. Shin. 2012. Oxidation resistance of iron and copper foils coated with reduced graphene oxide multilayers. ACS Nano 6: 7763–7769.

[13] Sahu, S. C., A. K. Samantara, M. Seth, S. Parwaiz, B. P. Singh, P. C. Rath and B. K. Jena. 2013. A facile electrochemical approach for development of highly corrosion protective coatings using graphene nanosheets. Electrochem. Commun. 32: 22–26.

[14] Singh, B. P., B. K. Jena, S. Bhattacharjee and L. Besra. 2013. Development of oxidation and corrosion resistance hydrophobic graphene oxide-polymer composite coating on copper. Surface and Coatings Technology 232: 475–481.

[15] Bunch, J. S., S. S. Verbridge, J. S. Alden, A. M. van der Zande, J. M. Parpia, H. G. Craighead and P. L. McEuen. 2008. Impermeable atomic membranes from graphene sheets. Nano Lett. 8: 2458–2462.

[16] Chen, S., L. Brown, M. Levendorf, W. Cai, S. -Y. Ju, J. Edgeworth, X. Li, C. W. Magnuson, A. Velamakanni, R. D. Piner, J. Kang, J. Park and R. S. Ruoff. 2011. Oxidation resistance of graphene-coated Cu and Cu/Ni alloy. ACS Nano 5: 1321–1327.

[17] Hu, J., Y. Ji, Y. Shi, F. Hui, H. Duan and M. Lanza. 2014. A review on the use of graphene as a protective coating against corrosion. Ann. J. Materials Sci. Eng. 1: 1–7.

[18] Li, X., W. Cai, L. Colombo and R. S. Ruoff. 2009. Evolution of graphene growth on Ni and Cu by carbon isotope labeling. Nano Lett. 9: 4268–4272.

[19] Dumée, L. F., L. He, Z. Wang, P. Sheath, J. Xiong, C. Feng, M. Y. Tan, F. She, M. Duke, S. Gray, A. Pacheco, P. Hodgson, M. Majumder and L. Kong. 2015. Growth of nano-textured graphene coatings across highly porous stainless steel supports towards corrosion resistant coatings. Carbon 87: 395–408.

CHAPTER 10

Advanced Micro/Nanocapsules for Self-healing Smart Anticorrosion Coatings

A Review of Recent Developments

*Kayla Lee,[1] Cynthia G. Cavazos,[1] Jacob Rouse,[1] Xin Wei,[1] Mei Li[2]
and Suying Wei[1],**

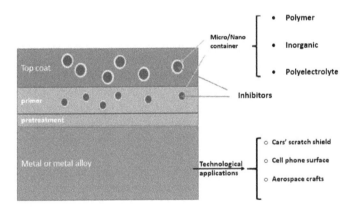

1. Introduction

Corrosion affects all industrial sectors where metal or metal alloys are used in their structures. Based on the study released by the U.S. Federal Highway Administration (FHWA) in 2012, the total annual estimated direct cost of corrosion in the U.S. is a staggering $276 billion—approximately 3.1% of the nation's Gross Domestic Product (GDP) [2]. The steel structure used for petrochemical pipelines at the gas

[1] Department of Chemistry and Biochemistry, Lamar University Beaumont, TX, USA.
[2] Department of Materials Science and Engineering, Qilu University of Technology, Jinan, China.
* Corresponding author: swei@lamar.edu

and oil field suffer from external as well as internal corrosion; failure of effective inspection and repairing may lead to fire and even explosion, which has caused fatal accidents in Texas, Ohio and others. Concrete and mortar may suffer from mechanical and weather forces and form cracks leading to costly maintenance and sometimes hazardous damages. Automotive industries benefit from corrosion inhibition by the protective coatings from surface damages due to sand or rain erosion. Aerospace industries surely need to consider corrosion protection for various components, and to be responsible for avoiding releasing toxic species into the space.

There are well developed mature strategies for preventing corrosion. In summary, cathodic protection is the conventional method to slow down the corrosion process. A sacrificial metal is used to convert all of the anodic sites, which is more easily corroded, on the protected metal surface to cathodic sites. The sacrificial metal is a more electronegative metal based on the galvanic series, e.g., iron materials are usually coated with a layer of zinc. However, the cathodic protection method requires a regular replacement of sacrificial metals. Another way to prevent corrosion is coating protection method. Different types of coatings have been well developed and widely used to protect corrosion. For this method, a multi-layered chemical barrier is employed to protect the underlying metal. The most important three layers are the pretreatment layer, the primer, and the topcoat [1]. The pre-treatment layer plays an important role to enhance the adhesion between the primer and the protected metal. The primer is the layer where the corrosion inhibitor is incorporated which could be galvanization metal elements or polymerizing agent. The top coat is applied to isolate the protected metal and other layers from the environmental factors, such as ultra-violet radiation, high external temperature, water, hot corrosive liquids, air pollution, acid rain, micro-organisms, etc. [3] However, the toxicity of the chemicals used in the multi-layered barrier, such as chromated pigments, epoxy resin, and polyurethane, is harmful to both the environment and human health [4]. Recently, graphene and its derivatives have been tested in replacement partially of the hazardous species in the coating layer, and showed promising corrosion protection results and potentially the alternative "greener" choices of coating materials [5].

The concept of self-healing refers to restoring the original entity of the structure autonomously. In the context of corrosion protection, it is hoped that internal or external damages could be repaired at the very beginning stage, e.g., micro-scale cracks, a process initiated by the mechanical stimuli imposed on the inhibitor. Self-healing coatings could be intrinsic or extrinsic, depending on the release mechanism of the anti-corrosion inhibitors. For intrinsic self-healing, the inhibitor is dispersed in the coating matrix, while for the extrinsic self-healing coating, the inhibitors are encapsulated or embedded in certain forms within the coating matrix. It requires an external stimulus to rupture the shell of the micro/nanocontainer, thus release the inhibitor to heal the corrosion [4], or an external stimulus to induce cross-linking or volume swelling, thus seal the damage from corrosion in wet and/or dry conditions [6, 7]. In this short chapter, we mainly focus on the extrinsic self-healing coatings for anti-corrosion applications. The key components in this type of self-healing coating are the micro/nanocapsules and the matrix, in which the anti-corrosion healing agent is typically encapsulated in the micro/nanocontainer as the core content. In recent years, there have been great efforts in developing multi-action self-healing coatings, in which more than one active material/inhibitor functioned for anti-corrosion purpose. In some cases, the second inhibitor is in the same core as the first one [8], while in other cases, the second inhibitor was anchored into the shell with the first kept in the core [9, 10]. These are so called smart dual inhibiting self-healing coatings.

In the following sections, new developments in micro/nanocontainers will be covered, namely polymer-, inorganic-, and polyelectrolyte-based shell materials. Additionally, for each type, we will also elaborate on the most studied core content, the shell rupture and self-healing mechanism, and current technological applications. At the end, concluding remarks and future perspective of self-healing coatings are provided.

2. Polymer-based Micro/Nano-Capsules

2.1 Developments in Polymeric Shell Materials

Commonly used microencapsulation techniques include *in situ* polymerization [11–13] and interfacial polymerization [11, 12, 14]; however, other methods include *in situ* condensation [12], solvent evaporation, dialysis, and supercritical fluid expansion, electrodeposition [15], or by using classical polymerization, including micro-emulsion, mini-emulsion, and surfactant-free emulsion. Shell materials that can be synthesized using the methods above include methyl cellulose (MC), gelatin· polyurea (PU), polyvinyl alcohol (PVA) [15], poly(urea-formaldehyde) (PUF) [11–13, 16–18], poly(melamine-formaldehyde) (PMF) [12], poly(melamine-urea-formaldehyde) (PMUF) [12, 16], double-walled polyurea, and double-walled polyurethane/poly(urea-formaldehyde) (PU/UF) [11].

Urea-formaldehyde microcapsules with linseed oil as the core materials were used for the healing of cracks in an epoxy coating by Suryanarayana et al. [18]. These microcapsules were synthesized via *in situ* polymerization in water/oil emulsion. Urea and formaldehyde materials react in an aqueous solution to form poly(urea-formaldehyde) (PUF). As molecular weight of PUF increases, the polar groups of this polymer decrease until PUF molecules become hydrophobic. After this process occurs, the hydrophobic molecules migrate to the surface of the oil/water droplets. The synthesized microcapsules were then applied to an epoxy coating. Similar containers were developed by interfacial polymerization of methylene diphenyl diisocyanate and polyamidoamine dendrimer [14].

Copper/liquid microcapsule composite coatings with polyvinyl alcohol, gelatin or methyl cellulose as shell materials were prepared by electrodeposition [15]. Gelatin and methyl cellulose as shell materials are easy to release and disperse quickly in composite coatings [16].

Yi et al. used a cost efficient and ecofriendly material, lignin as Pickering emulsion stabilizer and the active hydroxyl groups in the chemical structure of lignin have the potential to react with isocyanate groups in oil phase and reinforce the stability of emulsions. Isophorone diisocyanate (IPDI) was loaded in lignin nanoparticle-stabilized oil-in-water Pickering emulsion templates (Fig. 1) [12].

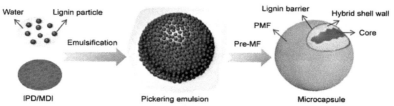

Fig. 1. Process of synthesis of multilayer composite microcapsules with core healing agents based on Pickering emulsion [12].

Es-haghi et al. prepared microcapsules of silane-treated ethyl cellulose microcapsule shell and used linseed oil as the core material. The three pre-silanes are 3-aminopropyl-trimethoxysilane (APS-EC), [3-(methacryloyloxy)-propyl]-trimethoxysilane (MPS-EC), and [3-(2,3-epoxypropoxy)-propyl]-trimethoxy-silane (EPS-EC). These microcapsules were prepared with solvent evaporation method. These silane-treated ethyl cellulose microcapsules in this study have the ability to increase the harmony of microcapsules with polymeric matrix in coating synthesis. These microcapsules would be an excellent fit for future studies of water-based self-healing coatings [19].

2.2 Shell Rupture and Healing Agent Sealing

In the low temperature experiment conducted by Kim et al., STP/DD-based self-healing coatings were applied to one side of steel panels. These panels were then stored at –20°C for 24 hours. Cross scratches were applied to the self-healing and control coatings while the panels were still within the chamber. When the coatings were removed from the chamber after 12 hours of –20°C temperatures, they were immersed in NaCl aqueous solution at –20°C. After 48 hours had gone by, the coatings were washed with distilled water and wiped dry, the coating surfaces were placed under observation by optical microscopy. All control samples in this experiment corroded, but STP/DD-based self-healing coatings showed no visual evidence of corrosion [20].

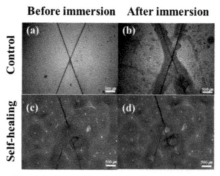

Fig. 2. Microscope images of scratched control and self-healing coated plates before and after immersion in sodium chloride solution. (a) Control before immersion; (b) control coating after immersion; (c) self-healing coating before immersion; (d) self-healing coating after immersion [20].

In an experiment by Suryanarayana et al., encapsulated linseed oil was released by a crack in the coating, the synthesized coating then filled the crack in a coating matrix. The linseed oil was then oxidized by oxygen in the atmosphere, which led to the finalization of solid film inside the crack [18].

In the experiment conducted by Vakhitov et al., containers containing two "green compounds", i.e., *N*-lauroylsarcosine (NLS) and linseed oil (LO), were incorporated into the waterborne acryl-styrene copolymer (ASC) coatings. The microcapsule shell consisted of polyvinyl alcohol (PVA) cross-linked by boric acid. Containers in the defected zone are also mechanically destroyed and the inhibitor is released quickly which immediately starts to heal the affected area. NLS forms a chelate complex. The strong adsorption of the material leads to a protective hydrophobic shield against the harmful impact of corrosive species on the metal surface, making this microcapsule combination an excellent contender for environments that are humid, near a body of water, or experience heavy rainfall [21].

Figure 3 below illustrates the healing concept of the microcapsules. A polymer shelled microcapsule suspended in a solution with the corresponding catalyst would be applied to a surface. When the surface experiences a crack or a rupture, the microcapsules would also experience this effect thereby exposing the core materials to the catalyst. As the catalyst and the core materials interact, polymerization occurs and bonds the crack closed. Part b of Fig. 3 shows a scanning electron microscope image of the fractured plane of a self-healing microcapsule with a ruptured PUF shell [22].

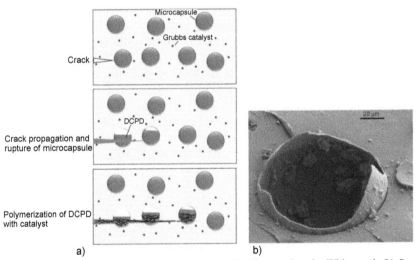

a)   b)

Fig. 3. (a) Representation of the functionality of self-healing coatings by White et al. (b) Scanning electron microscope image of a ruptured PUF microcapsule [22].

2.3 Core Materials

The core materials, or healing agents, for micro and nano-capsules should be able to react with multiple kinds of exposures after being ruptured including, the coating matrix, oxygen and UV light [11]. Healing agents include, but are not limited to, dicyclopentadiene (DCPD), organic solvents, epoxy resin, chlorobenzene, mercaptan, curing agents and drying oils. Multiple processing parameters such as agitation speed, temperature and emulsifier concentration influence properties such as morphology, shell wall parameters, size distribution, storage stability and core content [12].

Song et al. reported an experiment involving cinnamide moiety encapsulated via polydimethylsiloxane shells (CA-PDMS) that were prepared and used as a healing agent. CA-PDMS was also microencapsulated with a urea-formaldehyde polymer shell. Upon photo-irradiation, CAPDMS generated viscoelastic substances which have intrinsic self-healing capabilities when scribed [17].

In the absence of catalyst, Samadzadeh et al. demonstrated the self-healing ability of encapsulated tung oil in PUF shell. Thanawala et al. evaluated self-healing coatings based on linseed oil. Compatibility of these oils was observed due to the presence of unsaturated fatty acids in linseed oil and tung oil; with the presence of these molecules crosslinking occurs with oxygen to make a stable complex. On the other hand, coconut oil as a non-drying oil has only 6.5 wt% unsaturated fatty acids and for centuries, this oil has been used in paints and coatings; however, with the low percentage of fatty acids, linseed and tung oil would be more compatible for application [16].

Silanol-terminated polydimethylsiloxane (STP) as a healing agent and dibutyltin dilaurate (DD) as a catalyst were separately microencapsulated to prepare a dual-capsule self-healing coating. The reaction behavior of STP in the presence of DD and the release of STP and DD from ruptured microcapsules were studied and analyzed at $-20°C$. Kim et al. evaluated the self-healing performance of the coating at $-20°C$ using corrosion tests, electrochemical tests, and saline solution permeability tests. Figure 4 shows the structures of STP and DD [20].

The research conducted by Wang et al. aimed to study the influence of linseed oil healing agents encapsulated in PUF microcapsules on anticorrosive properties of epoxy coatings by testing different weight percentages of microcapsules in the coating matrix. By adding 10 wt% linseed oil encapsulated PUF microcapsules into epoxy coatings, water-uptake was increased; however, the linseed oil coating at microcrack will lose barrier properties within one day in a 3.5 wt% NaCl aqueous solution.

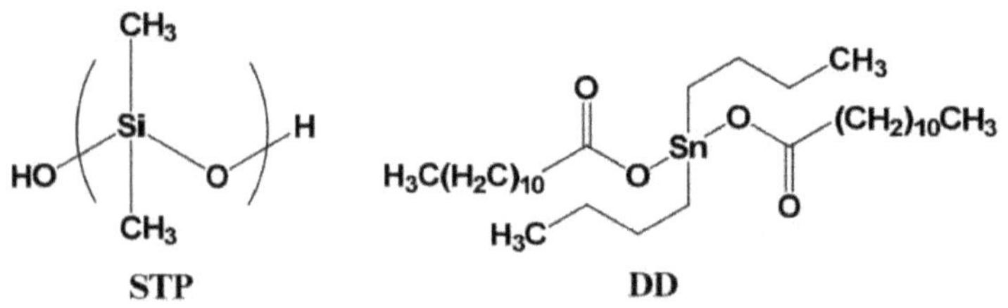

Fig. 4. Chemical structures of STP and DD used in the experiment by Kim et al. [20].

2.4 Technological Applications

Metals and metal alloys tend to corrode in their applying environment, and one of the most practical ways to prevent this from happening and thus save the cost of repairing or replacing is to use a coating [16].

Applications suitable for polymer micro/nano capsules include protecting inner surfaces of aircraft fuel tanks, transport lines, and reservoirs. Enhanced functionality was shown when smart coatings were applied to military equipment and industrial gas turbines. Self-cleaning coatings have been used in the automobile industry on car bodies, mirrors, and windshields. Nissan's 'Scratch Guard Coat' heals artificial scratches by a reflow resistant of abrasion that is five times more successful than traditional paints. These coatings can also be applied to sensors, lenses, cameras, and telescopes in the optical industry [23].

Applications have also been successful for anti-corrosion and non-stick anti-foulant purposes in marine, medical, dental and aerospace industries, coatings for windows, in the textile industry, solar modules, and exterior paints. Coatings have also been applied in cook wares (scratch prevention), building materials, and in electronics. Antibacterial coatings have been applied to naval structures to reduce losses due to microbial corrosion [23].

Self-healing micro and nano capsules show optimum performance at or above room temperature, but low temperature self-healing capabilities are also in high demand for cold climates. For aerospace applications, temperatures can be as low as –60°C; therefore, self-healing coatings with temperature tolerant capabilities would be excellent additions at these temperatures. Kim et al. reported self-healing coatings with temperature tolerance as low as –20°C; this coating was the first example of self-healing coatings at low temperatures without the need of an application of heat. In this experiment, silanol-terminated polydimethylsiloxane (STP) was used as a healing agent and dibutyltin dilaurate (DD) was used as a catalyst and these substances were separately microencapsulated to form dual-capsule self-healing coatings [20].

Finally, nanocontainers with polymer shells can be successfully applied to water-borne polymer coatings to be used for protection of aluminum alloys and steel. In most instances, these coatings can be easily dried in the open air and have advantages of high loading capacity and possibility to design permeability properties of the nanocontainer shell. The downfall of these nanocontainers is that their nature limits the applications suitable for them. For example, they are not a suitable choice for oil-borne coatings because of their potential solubility in the organic materials that make up the self-healing coating and they cannot withstand high temperatures and pressures [24].

3. Inorganic Based Micro/Nano-capsules

3.1 Mesoporous Silica as Nanocontainers

3.1.1 About the Material

Mesoporous silica is one of the most studied materials when it comes to active anti-corrosion coatings as it has many properties. For instance, mesoporous silica has a low toxicity [25], high surface area, large pore volume, and controllable stimulation [26]. These properties contribute to high loading capacity, and release of inhibitors, both of which are used in self-healing coatings [26]. As for synthesizing this material, the traditional way involves the formation of SiO_2 templates, then removing the templates through calcination or acid extraction, and sometimes functionalization of SiO_2, followed by loading inhibitors [26]. For example, to make a light-controlled nanocontainer, first there was the synthesis of SiO_2 particles, then mesoporous silica was synthesized with cetyltrimethyl ammonium bromide (CTAB) and azobenzene. The azobenzene is used for its isomerization property, which contributes to a light-controlled mechanism and CTAB as a template; finally, there is the removal of CTAB which creates the containers as seen in Fig. 5 [27]. Another common material used in the synthesis of these containers is the tetraethyl orthosilicate (TEOS). TEOS is to provide the silica source and more than often is used for the first step, the synthesis of SiO_2 [26]. For example, in one case the synthesis of mesoporous silica containers starts off with 0.21 g of NaOH being dissolved in 363 mL of deionized water at 80°C. Then 0.75 g of CTAB is added to the solution, followed by 3.75 mL of TEOS [26]. This process created the nanocontainers.

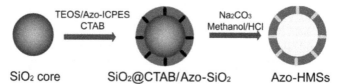

Fig. 5. A traditional way of formulating mesoporous silica nanocontainers synthesized with azobenzene [27].

3.1.2 Mechanism of Corrosion Inhibitor Release

The main use of nanocontainers is inhibitor loading and releasing. Using mesoporous silica as the nanocontainer provides high loading capacity and controllable stimulation of inhibitor release. The issue is that many inhibitor loading methods are complex; however, researchers found a one-step processes to load the container. Unlike the traditional way, the template, CTAB, remains in the container, because the SiO_2-CTAB is strongly affected by the pH in its surroundings. Letting the CTAB stay in the container makes it a one-step process [26]. As for the release of inhibitors, many mesoporous silica nanocontainers are built with pH sensitive release mechanisms, like the SiO_2-CTAB containers. Since the pH changes due to the cathodic and anodic reactions of corrosion, researchers have used the pH change to activate inhibitor release. The result is that the nanocontainers hold the inhibitors in neutral environments and release them when in contact with acidic or alkaline surroundings [25, 26, 28–30]. This means that the inhibitors will be released near the corrosion site. The release rate of the inhibitors varies in different cases. For example, the nanocontainer that is inhibited with molybdate releases more inhibitors at pH above 6 [30], whereas another container loaded with dodecylamine performs better at a lower pH [25].

3.1.3 Types of Inhibitors

The types of inhibitors in these nanocontainers are usually what makes the self-healing part of these coatings. Researchers have used different inhibitors for the nanocontainers including, 1H-benzotriazole (BTA) [26, 27], molybdate [28, 30], dodecylamine [25], and 8-hydroxyquinoline (8-HQ) [29]. Each have their own unique properties or combined with a certain material could enhance certain properties. For example, the high solubility of BTA combined with a large proportion of CTAB created a larger loading capacity for the nanocontainer. In addition to that, BTA is one of the most efficient and widely used inhibitors for the corrosion of copper and its alloys [26]. Molybdate is an inhibitor that within mesoporous silica has been proven to produce a better protected surface than pyrrole/mesoporous silica coating [28]. Dodecylamine has been proven to have self-healing abilities with scanning vibrating electrode technique (SVET). It is based on the formation of a film on the metal surface and blocks aggressive ions [25]; likewise, 8-HQ is chosen for its excellent corrosion inhibition [29].

3.1.4 Technological Applications

Mesoporous silica nanocontainers can be applied in many alloys and metals for anticorrosion. Some methods have been proven to have long lasting corrosion protection making it possible for long term anti corrosion applications. The one step method has the possibility to be extended to many inhibitors and templates with different response properties [26]. The inhibitors also have a part to play as some work well with certain metals, for example, the inhibitor BTA works well with copper and its alloys [26]. Anything that uses copper could work well with a coating loaded with this inhibitor. Coins or, more importantly, electrical wiring could use this coating. However, some of the methods mentioned are still in their early phase and only more research can tell where these coatings can be applied.

3.2 Two Types of Nanocontainers in One Coating

3.2.1 About the Material

Having more than one container in a coating is not as common, but there are researchers who have tried to combine a few containers to make an anticorrosion coating. While there is the use of mesoporous silica, there are more materials that can be used such as halloysite tubes [24, 31, 32] and titanium dioxide nanotubes (TNT) [33]. This section talks about the types of containers in one coating, specifically one with TNT and mesoporous silica. To synthesize the TNT, a hydrothermal method was applied. TiO_2 powder was mixed with NaOH solution and stirred. It is then heated at 130°C for 10 h. The resulting precipitate was washed in distilled water and HCl until a pH of 7 was obtained. Afterward, it was dried and readied to be loaded with pre-epoxy polymer. Then the synthesis of the mesoporous silica was started with the amphiphilic triblock copolymer being mixed into a solution of water and HCl. TEOS is added and together both of those formed a precipitate. To remove the triblock copolymer, the precipitate was washed with water and ethanol. It is dried and then calcined to decompose the triblock copolymer [33], thus creating the mesoporous silica particles.

3.2.2 Mechanism

The way this coating works is slightly different from the single mesoporous silica nanocontainer coating, since now it has two different containers each with a specific job. The first thing that has to happen is that the TNT has to open first in order for the epoxy pre-polymers to react with the amine curing agent. Because the coating is then immersed in an electrolyte solution, the TNT slowly starts releasing the epoxy pre-polymer. The polymer comes in contact with the amine curing agent and can begin the self-healing process. It was proven that this coating recovered 57% of anti-corrosive property with an electrochemical impedance spectroscopic (EIS) analysis. The test demonstrated that the impedance value of this coating continued to increase after 48 h making it ideal for long term protection against corrosion. The results are confirmed by the scanning electron microscope (SEM) images shown in Fig. 6, which compares the coating from day 1 to day 5, showing the nearly healed scratch in the metal [33].

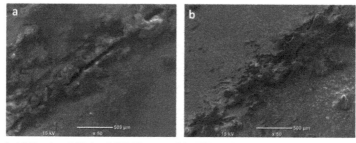

Fig. 6. (a) Day 1 and (b) day 5 SEM images of the TiO_2 and mesoporous silica containers [33].

3.2.3 Encapsulation

The encapsulation for this coating is different from the regular mesoporous silica containers. As mentioned, this coating has two types of containers in it, each with a different material in it. The TNT and mesoporous silica particles store epoxy pre-polymers and amine curing-agent, respectively. To load the TNT, a solution of Epon 815c/15 phr TNT epoxy resin is produced and placed in a vacuum jar. The pre-epoxy polymers are then able to replace the air inside the nanotubes as shown in Fig. 7a. To load the mesoporous silica, the particles were prepared with Epikure 3223 and then shaken gently to immobilize the amine molecules in the pores of mesoporous silica [33] as seen in Fig. 7b. Afterward, the Epon 815c/15 phr TNT is diluted with Epon 826 and then mixed with amine curing agent immobilized in mesoporous silica.

3.2.4 Technological Applications

This coating demonstrates slow and steady healing abilities. As a result of that, this coat could be appropriate for long-term anti-corrosion applications. Metal structures,

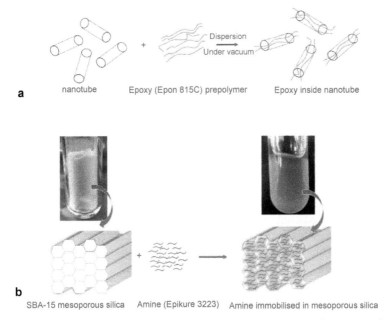

Fig. 7. (a) Representation of epoxy pre-polymer loaded into TiO$_2$ tubes and (b) the representation of amine molecules being loaded into mesoporous silica containers [33].

spacecraft, aircraft, and materials that will be around for a long time could benefit from this coating. On top of that, this coating has epoxy in it, making it compatible with epoxy composites [33]. It may only be in its beginnings, but it has much potential. The only downside is the amine curing agent, which is highly reactive and that makes it difficult to load [33]; however, researchers are still looking into this type of coating.

3.3 Layered Double Hydroxides as Containers

3.3.1 About the Material

In recent years, many researchers have looked into layered double hydroxides (LDH) as a potential material for anti-corrosion for metals. LDHs are a type of anionic clay [34, 35, 36, 37] that have unique properties such as ion exchange [34, 35, 37, 38, 39, 40, 41], ability to carry and release inhibitors [42], high corrosion resistance [37], self-healing ability [35, 37], and adjustable structures [35]. There are several ways to synthesize this material: *in situ, ex situ* or also known as two-step process [34, 35, 43] and anion-exchange [34, 37, 42]. The *in situ* process is where the LDH is directly grown on the metal substrate; however, this process makes fragile coatings [35]. In order to compensate for that, researchers have come up with the idea to treat it with plasma electrolytic oxide (PEO) to combine the self-healing abilities of LDH and barrier protection of PEO [40]. The two step or *ex situ* method first synthesizes the LDH precursor powder and the coating is then fabricated with a specific process [34, 43]. In the anion exchange method, small interlayer anions are synthesized with

LDH precursors and then in certain conditions, the anions exchange with interlayer anions of the precursors [34].

3.3.2 Mechanism

There are two possible ways for this coating to function. It is either with anion inhibitors being released or ion exchange [41]. For the ion exchange, the mechanism mostly depends on the pH of its surroundings [41, 43]. It is known that in high levels of pH, the LDH tends to dissolve slightly. This prompts the release of inhibitor anions, and the capture of the aggressive chloride ions. In addition, some metals, such as aluminum, could interact with the inhibitor ions and form an oxide layer for more protection [41]. This process can be seen in Fig. 8. There is also the anion exchange mechanism that reacts when corrosive ions are near. When these ions are near, the LDH can capture them and release the anodic inhibitors. There are instances when not just anodic inhibitors are released. For instance, Zn-Al-LDH release Zn^{2+} and Al^{3+} cations due to the solubility of LDH. These cations are cathodic inhibitors and could also block the cathodic and anodic areas through precipitation by reacting with phosphate [42].

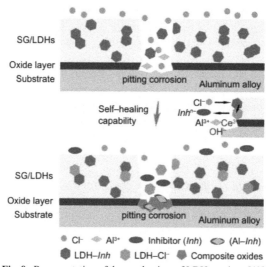

Fig. 8. Representation of the mechanism of LDH coatings [42].

3.3.3 Types of Inhibitors

Because LDHs have layers, different types of inhibitors, or in some cases anions, can be put in them. Some types of inhibitors are molybdate [41], vanadate [36, 39, 41, 44], phosphate [42], zinc [42], tungstate [35], 5-methyl-1,3,4 thiadiazole-2-thiol (MTT) [38] and BTA [40]. Most of the inhibitors are anions because not only are they compatible with the LDH, but also many researchers use the anion exchange to load them in the containers. In some cases, they mix the inhibitors such as molybdate and vanadate in order to enhance the active anti-corrosion properties of LDH [41], but for the most part only one inhibitor is used.

3.3.4 Technological Applications

This coating has been looked into by many researchers and for the most part these coatings can protect metals enough to be used in anticorrosion applications. LDH coatings have been tried on magnesium alloys, aluminum alloys, and steel [43]. As far as applications go, coatings can be applied to materials made out of these metals; however, not all work as well as they should. For example, magnesium is practically non-reactive or passive when put in alkaline solutions [35]. LDH depends on the alkaline solutions around it so that it can self-heal damaged metal. Despite that, many researchers are still exploring the possibilities of LDH.

3.4 Cation Exchange Clays

3.4.1 About the Material

Cation-exchange clays are an area that has been looked into by many researchers. These clays are significant because they can release inhibitors and have the ability to take aggressive ions [45]. One such clay is bentonite, also known as montmorillonite. These clays show high cation exchange abilities, which is the basis of the whole system of this coat. The materials used to formulate this coating are polyvinyl-butyral (PVB) and bentonites with alkaline earth cations such as Mg^{2+}, Ca^{2+}, Sr^{2+}, and Ba^{2+} [45, 46]. The bentonite could also be synthesized with rare earth cations like Y^{3+} and Ce^{3+}; however, the rare earth cations are not as effective as alkaline earth cations [46]. In order to make the material, the first thing is to synthesize PVB with the bentonite clay. The next thing is to get the alkaline and earth cations in the bentonite by exchanging the bentonite in aqueous solutions of alkaline chloride or rare earth chlorides [45]. After all that is done, to make sure there is no more sign of chloride ions in the bentonite, they are washed by centrifugation and re-dispersion in distilled water multiple times. Finally, they are washed in ethanol and dried for four hours [45].

3.4.2 Self-healing Mechanism

The main mechanism of this type of coating is based on the cation exchange mechanism. This mechanism prevents cathodic disbondment, which in turn prevents cathodic oxygen reduction, the main cause of corrosion. In order for the cation exchange mechanism to happen, the coating is synthesized and put on a metal substrate. Then an electrolyte is put directly on the metal through a cut in the coating. This electrolyte creates a layer and initiates delamination between the coating and the metal [45]. The cations are then able to exchange when aggressive ions are trapped and the counter-ions are released. Once they are exchanged, Ca^{2+} and OH^-, react to form a calcium hydroxide product. It is this insoluble product that plays the main role of preventing cathodic disbondment. The product blocks the cathodic site of the metal substrate [46] and reduces the electrolyte conductivity [45], thus preventing cathodic disbondment from happening.

3.4.3 Technological Applications

Although, cation exchange is a promising way to slow down cathodic disbondment, there is an issue. The environment in which the coating is put in can affect how well it works. It has been observed that a calcium-exchange bentonite is not enough to protect the surface of the metal over a long period of time when put against chloride ions [46]. Hence, applications for this method of anti-corrosion may be limited in order for this method to work properly. On the other hand, some calcium-exchange bentonites in a different environment have been proven to compete with chromate coatings, making it possible to replace those with these coatings [45].

4. Polyelectrolyte Multilayer Based Micro/Nano-capsules

4.1 Polyelectrolyte Nanocontainers Composition and Production Methods

Polyelectrolyte multilayer nanocontainers are a protective barrier that contains different solutions that can be used to heal damage caused by corrosion. Nanocontainers contain various inhibitors and are meant to open when certain circumstances are met. Polyelectrolyte nanocontainers are used inside coatings such as paint, resin, or other materials such as zinc. All the nanocontainers compositions that will be mentioned are used to contain solvents that are meant to heal damage caused to metals. Polyelectrolyte nanocontainers are most commonly created by a method called Layer-by-Layer. The Layer-by-Layer process has been around since the 1990s and is characterized by the iterative absorption of oppositely charged nanoparticles on a template particle or surface (Fig. 9) [24]. Benefits of a layer-by-layer approach when creating nanocontainers is the control it allows over composition and thickness. Layer-by-layer assembly allows the use of polyelectrolytes to be layered with themselves and other types of particles which can include any charged material. Charged materials include nanoparticles, lipids, viruses, and enzymes [24, 47]. Observation has concluded that when weak to weak electrolytes or strong to weak electrolytes are paired, there is greater anti-corrosion protection than strong to strong electrolytes being paired [48]. Poly acrylic acid is a weak polyelectrolyte that has a degree of dissociation between its carboxyl groups that decrease at lower pH, which makes this polyelectrolyte useful since corrosion generally is also associated with a small-scale pH change around the area affected. Nanocontainers are formed from this substance by dropping poly acrylic acid on to a particles surface, then removing any excess amount. Once the excess is removed, the solution meant to be contained is added. Once all the material that will be contained is contained the excess is removed. Lastly, extra layers of poly acrylic acid are added [17]. Polydiallydimethyl ammonium chloride (PDADMAC) was also used; polyelectrolyte nanocapsules created with PDADMAC have the potential advantage of not being reliant on the pH of the medium used [50]. When kept at a small size, this nanocontainer creates no significant negatives for an epoxy-based coatings barrier. Nanocontainers made from this nanocontainer are produced in a similar manner as poly acrylic acid. PDADMAC can also be used to make nanocontainers with poly sodium 4-strenesulfonte (PSS) when used as a polycation-polyanion pair [51]. In Fig. 10, PAA and PDADMAC

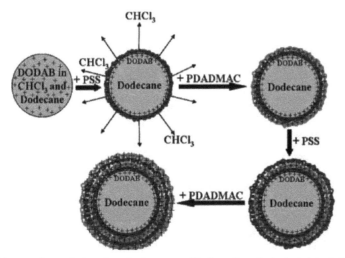

Fig. 9. Diagram of steps during a layer-by-layer assembly formation of polyelectrolyte shell [24, 53].

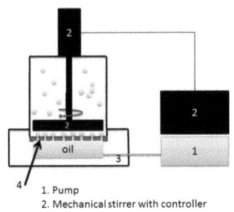

1. Pump
2. Mechanical stirrer with controller
3. Membrane holder
4. Membrane

Fig. 10. Diagram of the set-up to create polyelectrolyte multilayer nanocontainers with a liquid core [52].

can be used together to create polyelectrolyte multilayer nanocontainers. Starting with PAA is negatively charged, followed by centrifugation to remove any loosely absorbed material. The particles then get dispersed through sonification. Next the PAA gets loaded with the substance desired to be encapsulated. Finally, the positively charged PDAMAC is added to finish the nanocontainer. This process is referred to as layer-by-layer absorption. If this coating is created with the use of benzotriazole (BTA) in mind, then creating it with hematite particles and salt can enhance the content of BTA by 50% [52]. Poly- allylamine hydrochloride (PAH) can be used with PSS to create nanocontainers. This process is done by absorption of positively charged PAH ions into SiO_2 followed by PSS and repeated centrifugation. After the desired number of layers were created, the SiO_2 was removed with hydrofluoric acid, leaving an empty container. A major issue with most polyelectrolyte multilayer based nanocontainers is balancing their size with their inhibitor load [24, 52]. Polyelectrolyte nanocontainers often incorporate other materials into them to create the shell when being produced. Two such materials are hematite and kaolinite. Both were used with poly acrylic acid and polydiallyldimethyl ammonium [49].

4.2 Breaking Mechanism for Polyelectrolyte Nanocontainers

Nanocontainers are designed to break open when specific forms of stimuli are obtained. These conditions are what make them useable inside coatings to protect various forms of metal from harm. A universal breaking mechanism is increased ionic strength of solutions exposed to a polyelectrolyte nanocontainer [54]. For polyelectrolyte multilayer nanocontainers, one breaking mechanism is ion exchange. Ion exchange occurs when an ion penetrates the coating [24, 30, 52]. One such example is with double hydroxide. Double hydroxide, along with other polyelectrolytes, reacts to negative chlorine ions when they are introduced. These ions can be found in salt water [52, 30]. Even though this is a method to break open nanocontainers, high concentrations of salt water can create an issue. High concentrations of salt water might end up breaking down protective layers too fast depending on the type of inhibitor located inside the nanocontainer. This only applies when constantly exposed to the material [52]. A change in pH is another breaking mechanism for polyelectrolyte multilayer nanocontainers. When corrosion starts to take place, a pH change occurs near the metals surface. The pH change in return reacts with the polyelectrolyte nanocontainer, since they are made from electrolytes that can easily be broken down when environmental pH shifts [50]. Along with the opening of polyelectrolyte nanocontainers due to pH, they can also be sealed when the pH returns to normal conditions again. pH change works best for nanocontainers composed of weak polyelectrolytes. Non-polar solvents can damage the integrity of polyelectrolyte shells, thus causing the container to open in response. Temperature increase works on shells that contain strong polyelectrolytes, but not weak. There are other ways to open polyelectrolyte nanocontainers, but they depend on having either specific electrolytes or other components incorporated inside the polyelectrolyte multilayer shell. Light (photothermal) can break down shells composed of light

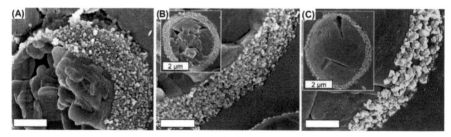

Fig. 11. Cryo-SEM visualizations showing that the regularity of particles on an oil-water interface is based on the degree of dissociation of corresponding polyelectrolytes. The polyelectrolytes used are silica-poly(allylamine hydrochloride) and the pH values of each emulsion are 8.5, 9.1, and 9.8, respectively. Unlabeled scale bars are 500 nm [24, 55].

sensitive elements. A magnetic treatment can break open a shell that contains magnetic particles. Oxidation or reduction can break open shells that contain redox materials. Notable redox materials that can be used are conductive polymers. Enzymatic degradation will open capsules with biodegradable components, but it is irreversible [24].

4.3 Inhibitors Inside Polyelectrolyte Multilayer Nanocontainers

Polyelectrolyte multilayer nanocontainers for self-healing anticorrosion coatings contain inhibitors meant to heal or prevent any corrosion caused to the surface of a material, in this case metals. They must be compatible with the type of nanocontainer used, in this case polyelectrolyte multilayer. The inhibitors also benefit more if they can prevent or stop a type of corrosion that correlates to how the nanocontainers are broken open. These inhibitors are the main agent that prevent corrosion from taking place by patching up any holes in the coating caused by chemical or mechanical damage. One inhibitor is benzotriazole (BTA). Benzotriazole is used as a corrosion inhibitor for metal and metal alloys including copper, aluminum, iron, and zinc. When released, it creates a passive layer that prevents corrosion. Benzotriazole is useful for conditions that offer external andic polarization. A major drawback of BTA is that it is water-soluble, but when used with polyelectrolyte nanocontainers inside a coating this drawback is lessened [49]. Benzotriazole is used as an inhibitor since it is great at covering any openings the coating may obtain due to damage, thus making it self-healing. The effectiveness of benzotriazole varies immensely with different concentrations, sizes, and layouts of the nanocontainers encapsulating it. Large mechanical damage may cause issues with coatings that rely only on BTA, since BTA may not be able to cover all the damage. Large scale damage is also an issue with many inhibitors since there is only so much an inhibitor can cover. BTA has

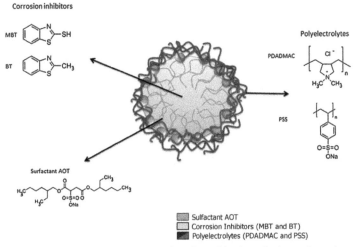

Fig. 12. Diagram of a polyelectrolyte nanocontainer formed by layer-by-layer absorption using poly sodium 4-strenesulfonte (PSS) and poly diallyldimethly ammonium (PDADMAC). The nanocontainer shown contains the corrosion inhibitor 2-mercatobenzothiazole (MBT) [51].

been tested to work well for prolonged periods in a corrosive environment when used with a well-made nanocontainer and coating [54]. 2-mercatobenzothiazole (MBT) was tested as a potential inhibitor. MBT self-heals by forming a thin film on an alloys surface protects the surface against corrosion. MBT is notably a good inhibitor for aluminum alloys. MBT is not an effective anti-corrosive inhibitor when used outside of nanocontainers. MBT does not affect adhesiveness of coatings that are using it. 2-methylbenzothiazole and 2-mercaptobenzothiazole work well against corrosion caused by small mechanical damage. Both do not work well against constant salt sprays as it can be possible for them to remove any new barrier formed. This was tested with a salt spray test and every case showed no inhibition to corrosion and received a macro rating of 5 [51].

4.4 Applications for Polyelectrolyte Nanocontainer Self-healing Anti-corrosion Coatings

Polyelectrolyte multilayer nanocontainers for self-healing coatings have numerous applications. If they can contain an inhibitor to protect the material, then they can be used for the material and most of its functions. Ways to prevent corrosion are desired because corrosion consumes 3% of the world's gross domestic product annually [50]. Applications for polyelectrolyte multilayer nanocontainer self-healing anticorrosion coatings are a way to potentially lower the losses. Generally, polyelectrolyte nanocontainers are added to coatings that already contain anti-corrosive properties. An example would be zinc coatings that are already known to prevent rust for various metals. Galvanized steel, which is steel that contains a coating made of zinc, is an effective use for polyelectrolyte multilayer nanocontainer self-healing anti-corrosion coatings. This makes these coatings useful also for construction projects such as fences, building frames, ductwork, and handrails. Galvanized steel is also often used to produce ladders. Besides steel, iron can also be galvanized [56]. Polyelectrolyte nanocontainers can also be used for steel without the coating being composed of zinc [57]. Another reason they are useable for zinc coatings on their own is because inhibitors such as BTA are an inhibitor for corrosion to zinc itself. Steel can easily rust since it is composed of iron, so coatings that can withstand or even self-heal would be a vital asset to production that uses these materials. Polyelectrolyte multilayer nanocontainer containing coatings also find purpose for metals such as copper, aluminum, and iron [49, 56], all of which can be used for building structures. Copper is known for its excellent electrical conductivity, making it useful for wiring. They can be freely used on any structure that isn't completely submerged under ocean water and is not mechanically damaged on a large scale. If the mechanical damage stays limited to that of small scratches and breaks, the coating should be effective. Polyelectrolyte multilayer nanocontainers that also include mesoporous silica have been used to create counter fire-based damage [58]. Polyelectrolyte nanocontainer self-healing anti-corrosion coatings can be used for aluminum alloys which are commonly applied to transportation and aerospace industry, since aluminum alloys have a low weight and excellent mechanical properties [51]. Overall polyelectrolyte multilayer nanocontainer containing self-healing electrolytes are an effective method to fight corrosion with a few concerns, but the outlook for their future in many applications is bright [24].

5. Concluding Remarks

Self-healing has seen some significant advancements in both capsule shell materials developments and the functionalities in the mechanistic self-healing response, namely dual (or multi-) smart self-healing coatings. There are many fields that may take advantage of the self-healing technology, and potentially bring economic and environmental benefits to the society. Though most studies are still limited to lab scale testing, there is great promise for upscale and commercialization of self-healing coatings. At a self-healing corrosion workshop called MUST in 2010, almost half the participants were from industries, which might be an indicator of tremendous interest in furthering the progress of these products. In fact, it has been several years since Nissan first released a self-healing product called "scratch shield", which is also applied to mobile phones. The company Akzo also announced a clear coating recently.

In addition to efficacy and cost consideration for the application of self-healing anti-corrosion coatings, environmental sustainability should also be considered in terms of choice of shell/core compositional/inhibitor materials and the related chemical processes. More non-organic based coating matrix might be a better option, considering delamination and releasing of the coating into the surroundings over its life cycle. Meanwhile, it does seem a great idea and trend to incorporate multi-functions into the self-healing coating system, generating a synergy of anticorrosion and micro/nanocapsule shell/core, and improve the synergy of self-healing and other secondary (or even ternary) functions [59]. Lastly, it will be highly desirable to have more accurate and quantitative simulation for the design and functioning of the self-healing coatings [60].

6. Acknowledgments

The authors would like to thank the Welch Foundation (v-0004) and the Mason Fund for partial financial support of the students (KL, CGC, and JR) involved in the book chapter writing.

References

[1] Wei, H., Y. Wang, J. Guo, N. Shen, D. Jiang, X. Zhang, X. Yan, J. Zhu, Q. Wang, L. Shu, H. Lin, S. Wei and Z. Guo. 2015. Advanced micro/nanocapsules for self-healing smart anticorrosion coatings. J. Mater. Chem. A3: 469–480.

[2] Koch, G. H. et al. 2002. Corrosion Costs and Preventive Strategies in the United States. Federal Highway Administration.

[3] Lutz, A., O. van den Berg, J. Wielant, I. De Graeve and H. Terryn. 2016. A multiple-action self-healing coating. Front. Mater. 2: 73.

[4] An, S., W. Lee, A. L. Yarin and S. S. Yoon. 2018. A review on corrosion-protective extrinsic self-healing: Comparison of microcapsule-based systems and those based on core-shell vascular networks. Chem. Eng. J. 344: 206–220.

[5] Physics World on feature materials by IOP publishing, April 2019.

[6] Saini, V., M. Tapavicza, C. Eloo, K. Braesch, H. Wack, A. Nellesen, A. M. Schmidt and S. J. Garcíac. 2018. Superabsorbent polymer additives for repeated barrier restoration of damaged powder coatings under wet-dry cycles: A proof-of-concept. Prog. Org. Coat. 122: 129–137.

[7] Yabuki, A., S. Tanabe and I. Fathona. 2018. Self-healing polymer coating with the microfibers of superabsorbent polymers provides corrosion inhibition in carbon steel. Surf. Coat. Tech. 341: 71–77.

[8] Siva, T. and S. Sathiyanarayanan. 2015. Self-healing coatings containing dual active agent loaded urea formaldehyde (UF) microcapsules. Prog. Org. Coat. 82: 57–67.

[9] Carneiro, J., A. F. Caetano, A. Kuznetsova, F. Maia, A. N. Salak, J. Tedim, N. Scharnagl, M. L. Zheludkevich and M. G. S. Ferreira. 2015. Polyelectrolyte-modified layered double hydroxide nanocontainers as vehicles for combined inhibitors. RSC Adv. 5: 39916–39929.

[10] Shi, H., F. Liu and E. H. Han. 2015. Surface-engineered microcapsules by layer-by-layer assembling for entrapment of corrosion inhibitor. J. Mater. Sci. Technol. 31: 512–516.

[11] Lang, S. and Q. Zhou. 2017. Synthesis and characterization of poly(urea-formaldehyde) microcapsules containing linseed oil for self-healing coating development. Prog. Org. Coat. 105: 99–110.

[12] Vijayan, P. and M. A. Al-Maadeed. 2016. 'Containers' for self-healing epoxy composites and coating: Trends and advances. Express. Polym. Lett. 10: 506–524.

[13] Wang, H. and Q. Zhou. 2018. Evaluation and failure analysis of linseed oil encapsulated self-healing anticorrosive coating. Prog. Org. Coat. 118: 108–115.

[14] Tatiya, P. D., R. K. Hedaoo, P. P. Mahulikar and V. V. Gite. 2013. Novel polyurea microcapsules using dendritic functional monomer: synthesis, characterization, and its use in self-healing and anticorrosive polyurethane coatings. Ind. Eng. Chem. Res. 52: 1562–1570.

[15] Xu, X. Q., Y. H. Guo, W. P. Li and L. Q. Zhu. 2011. Electrochemical behavior of different shelled microcapsule composite copper coatings. Int. J. Min. Met. Mater. 18: 377–384.

[16] Ataei, S., S. N. Khorasani, R. Torkaman, R. E. Neisiany and M. S. Koochaki. 2018. Self-healing performance of an epoxy coating containing microencapsulated alkyd resin based on coconut oil. Prog. Org. Coat. 120: 160–166.

[17] Song, Y. K. and C. M. Chung. 2013. Repeatable self-healing of a microcapsule-type protective coating. Polym. Chem-UK. 4: 4940.

[18] Suryanarayana, C., K. C. Rao and D. Kumar. 2008. Preparation and characterization of microcapsules containing linseed oil and its use in self-healing coatings. Prog. Org. Coat. 63: 72–78.

[19] Es-haghi, H., S. M. Mirabedini, M. Imani and R. R. Farnood. 2014. Preparation and characterization of pre-silane modified ethyl cellulose-based microcapsules containing linseed oil. Colloid. Surface. A 447: 71–80.

[20] Kim, D. M., Y. J. Cho, J. Y. Choi, B. J. Kim, S. W. Jin and C. M. Chung. 2017. Low-temperature self-healing of a microcapsule-type protective coating. Materials 10: 1079.

[21] Vakhitov, T. R., V. E. Katnov, P. V. Grishin, S. N. Stepin and D. O. Grigoriev. 2017. Biofriendly nanocomposite containers with inhibition properties for the protection of metallic surfaces. Proc. R. Soc. A 473: 20160827.

[22] White, S. R., N. R. Sottos, P. H. Geubelle, J. S. Moore, M. R. Kessler, S. R. Sriram, E. N. Brown and S. Viswanatha. 2001. Autonomic healing of polymer composites. Nature 409: 794–797.

[23] Ulaeto, S. B., R. Rajan, J. K. Pancrecious, T. P. D. Rajan and B. C. Pai. 2017. Developments in smart anticorrosive coatings with multifunctional characteristics. Prog. Org. Coat. 111: 294–314.

[24] Grigoriev, D., E. Shchukina and D. G. Shchukin. 2017. Nanocontainers for self-healing coatings. Adv. Mater. Interfaces 4: 1600318.

[25] Falcon, J. M., L. M. Otubo and I. V. Aoki. 2016. Highly ordered mesoporous silica loaded with dodecylamine for smart anticorrosion coatings. Surf. Coat. Tech. 303: 319–329.

[26] Xu, J., Y. Cao, L. Fang and J. Hu. 2018. A one step preparation of inhibitor-loaded silica nanocontainers for self-healing coatings. Corros. Sci. 140: 349–362.

[27] Chen, T., R. Chen, Z. Jin and J. Liu. 2015. Engineering hollow mesoporous silica nanocontainers with molecular switches for continuous self-healing anticorrosion coating. J. Mater. Chem. A 3(18): 9510–9516.

[28] Keyvani, A., M. Yeganeh and H. Rezaeyan. 2017. Application of mesoporous silica nanocontainers as an intelligent host of molybdate corrosion inhibitor embedded in the epoxy coated steel. Pro. Nat. Sci-Mater. 27(2): 261–267.

[29] Shi, H., L. Wu, J. Wang, F. Liu and E. Han. 2017. Sub-micrometer mesoporous silica containers for active protective coatings on AA 2024-T3. Corros. Sci. 127: 230–239.

[30] Yeganeh, M., S. M. Marashi and N. Mohammadi. 2018. Smart corrosion inhibition of mild steel using mesoporous silica nanocontainers loaded with molybdate. J. Nanosci. Nanotechnol. 14: 143–151.

[31] Shchukina, E., D. Shchukin and D. Grigoriev. 2017. Effect of inhibitor-loaded halloysites and mesoporous silica nanocontainers on corrosion protection of powder coatings. Prog. Org. Coat. 102: 60–65.

[32] Shchukina, E., D. Shchukin and D. Grigoriev. 2018. Halloysites and mesoporous silica as inhibitor nanocontainers for feedback active powder coatings. Prog. Org. Coat. 123: 384–389.

[33] Vijayan, P. and M. A. S. A. Al-Maadeed. 2016. TiO$_2$ nanotubes and mesoporous silica as containers in self-healing epoxy coatings. Sci. Rep. 6: 38812.

[34] Guo, L., W. Wu, Y. Zhou, F. Zhang, R. Zeng and J. Zeng. 2018. Layered double hydroxide coatings on magnesium alloys: a review. J. Mater. Sci. Technol. 34: 1455–1466.

[35] Zeng, R., X. Li, Z. Liu, F. Zhang, S. Li and H. Cui. 2015. Corrosion resistance of Zn-Al layered double hydroxide/poly(lactic acid) composite coating on magnesium alloy AZ31. Front. Mater. Sci. 9: 355–365.

[36] Zhang, C., X. Luo, X. Pan, L. Liao, X. Wu and Y. Liu. 2017. Self-healing Li-Al layered double hydroxide conversion coating modified with aspartic acid for 6N01 Al alloy. Appl. Surf. Sci. 394: 275–281.

[37] Zhang, F., C. Zhang, R. Zeng, L. Song, L. Guo and X. Huang. 2016. Corrosion resistance of the superhydrophobic Mg(OH)$_2$/Mg-Al layered double hydroxide coatings on magnesium alloys. Metals 6: 85.

[38] Liu, A., H. Tian, W. Li, W. Wang, X. Gao, P. Han and R. Ding. 2018. Delamination and self-assembly of layered double hydroxides for enhanced loading capacity and corrosion protection performance. Appl. Surf. Sci. 462: 175–186.

[39] Yasakau, K. A., A. Kuznetsova, S. Kallip, M. Starykevich, J. Tedim, M. G. S. Ferreira and M. L. Zheludkevich. 2018. A novel bilayer system comprising LDH conversion layer and sol-gel coating for active corrosion protection of AA2024. Corros. Sci. 143: 299–313.

[40] Zhang, G., L. Wu, A. Tang, Y. Ma, G. Song, D. Zheng, B. Jiang, A. Atrens and F. Pan. 2018. Active corrosion protection by a smart coating based on a MgAl-layered double hydroxide on a cerium-modified plasma electrolytic oxidation coating on Mg alloy AZ31. Corros. Sci. 139: 370–382.

[41] Zhang, Y., P. Yu, J. Wu, F. Chen, Y. Li, Y. Zhang, Y. Zuo and Y. Qi. 2018. Enhancement of anticorrosion protection via inhibitor loaded ZnAlCe-LDH nanocontainers embedded in sol-gel coatings. J. Coat. Technol. Res. 15(2): 303–313.

[42] Alibakhshi, E., E. Ghasemi, M. Mahdavian, B. Ramezanzadeh and S. Farashi. 2016. Fabrication and characterization of PO$_4^{3-}$ intercalated Zn-Al-layered double hydroxide nanocontainer. J. Electrochem. 163: C495–C505.

[43] Zhang, F., C. Zhang, L. Song, R. Zeng, Z. Liu and H. Cui. 2015. Corrosion of *in-situ* grown MgAl-LDH coating on aluminum alloy. T. Nonferr. Metal. Soc. 25: 3498–3504.

[44] Serdechnova, M., M. Mohedano, B. Kuznetsov, C. L. Mendis, M. Starykevich, S. Karpushenkov, J. Tedim, M. G. S. Ferreira, C. Blawert and M. L. Zheludkevich. 2017. PEO coatings with active protection based on *in-situ* formed LDH nanocontainers. J. Electrochem. Soc. 164: C36–C45.

[45] Williams, G. and H. N. McMurray. 2017. Inhibition of corrosion driven delamination on iron by smart-release bentonite cation-exchange pigments studied using a scanning kelvin probe technique. Prog. Org. Coat. 102: 18–28.

[46] Vega, J. M., N. Granizo, J. Simancas, I. Diaz, M. Morcillo and D. de la Fuente. 2017. Exploring the corrosion inhibition of aluminum by coatings formulated with calcium exchange bentonite. Prog. Org. Coat. 111: 273–282.

[47] Gu, Y., X. Huang, C. G. Wiener, B. D. Vogt and N. S. Zacharia. 2015. Large-scale solvent driven actuation of polyelectrolyte multilayers based on modulation of dynamic secondary interactions. ACS Appl. Mater. Inter. 7: 1848–1858.

[48] Fan, F., C. Zhou, X. Wang and J. Szpunar. 2015. Layer-by-layer assembly of a self-healing anticorrosion coating on magnesium alloys. ACS Appl. Mater. Int. 7: 27271–27278.

[49] Kamburova, K., N. Boshkova, N. Boshkov, G. Atanassova and T. Radeva. 2018. Hybrid zinc coatings for corrosion protection of steel using polyelectrolyte nanocontainers loaded with benzotriazole. Colloid. Surface. A 559: 243–250.

[50] Qin, S., Y. Cubides, S. Lazar, R. Ly, Y. Song, J. Gerringer, H. Castaneda and J. C. Grunlan. 2018. Ultrathin transparent nanobrick wall anticorrosion coatings. ACS Appl. Nano. Mater. 1: 5516–5523.

[51] Kopeć, M., K. Szczepanowicz, G. Mordarski, K. Podgórna, R. P Socha, P. Nowak, P. Warsynski and T. Hack. 2015. Self-healing epoxy coatings loaded with inhibitor-containing polyelectrolyte nanocapsules. Prog. Org. Coat. 84: 97–106.

[52] Kopeć, M., K. Szczepanowicz, G. Mordarski, K. Podgórna, R. P. Socha, P. Nowak, P. Warsynski and T. Hack. 2016. Liquid-core polyelectrolyte nanocapsules produced by membrane emulsification as carriers for corrosion inhibitors. Colloid. Surface. A 510: 2–10.

[53] Grigoriev, D. O., T. Bukreeva, H. Möhwald and D. G. Shchukin. 2008. New method for fabrication of loaded micro- and nanocontainers: emulsion encapsulation by polyelectrolyte layer-by-layer deposition on the liquid core. Langmuir. 24: 999–1004.

[54] Li, Z., B. Qin, X. Zhang, K. Wang, Y. Wai and Y. ji. 2015. Self-healing anti-corrosion coatings based on polymers of intrinsic microporosity for the protection of aluminum alloy. RSC Adv. 5: 104451–104457.

[55] Haase, M. F., D. Grigoriev, H. Moehwald, B. Tiersch and D. Shchukin. 2011. Nanoparticle modification by weak polyelectrolytes for pH-sensitive Pickering emulsions. Langmuir. 27: 74–82.

[56] Kamburova, K., N. Boshkova, N. Boshkov and T. Radeva. 2016. Design of polymeric core-shell nanocontainers impregnated with benzotriazole for active corrosion protection of galvanized steel. Colloid. Surface. A 499: 24–30.

[57] Syed, J. A., S. Tang, H. Lu and X. Meng. 2015. Smart PDDA/PAA multilayer coatings with enhanced stimuli responsive self-healing and anti-corrosion ability. Colloids. Surfaces. A 476: 48–56.

[58] Jiang, S., G. Tang, J. Chen, Z. Huang and Y. Hu. 2018. Biobased polyelectrolyte multilayer-coated hollow mesoporous silica as a green flame retardant for epoxy resin. J. Hazard. Mater. 342: 689–697.

[59] Huang, Y., L. Deng, P. Ju, L. Huang, H. Qian, D. Zhang, X. Li, H. A. Terryn and J. M. C. Mol. 2018. Triple-action self-healing protective coatings based on shape memory polymers containing dual-function microspheres. ACS Appl. Mater. Interfaces 10: 23369–23379.

[60] Javierre, E. 2019. Modeling self-healing mechanisms in coatings: approaches and perspectives. Coatings 9: 122.

CHAPTER 11

Plasma Electrolytic Oxidation
Anticorrosive and Biocompatible Coatings

Aleksandra A. Gladkova,[1,] Dmitriy G. Tagabilev[2] and Miki Hiroyuki[3]*

1. Introduction

Recently, much attention has been paid to the development and modernization of surface treatment methods, with the fields of detail application and its resistance to the environmental exposure being mostly determined by the surface properties. Prevention of material damage by coatings has the potential of saving resources, such as operating costs or decreasing equipment depreciation expenses. For years, many countries have been looking for ways to make products that are both safe for the environment and reduce or eliminate any health concerns for their employees. Considerable collaborative work has been done to find environmentally compliant substitutes for chromium (Cr), in particular hexavalent chromium, as these coatings can successfully replace non-environmentally friendly technologies for corrosion protection [1–3]. Moreover, treated surface layers provide the ability to modify substrate surfaces to provide better biocompatibility, or control the rate of biomaterial degradation.

Thanks to the achievements of modern medicine, life expectancy continues to grow. Patients' demands for a quality of life are increasing; yet at the same time, the number of people with musculoskeletal system injuries is also steadily increasing every year. In addition, with the increase in life expectancy, the number of people who have indications for joint replacement as a result of long-lasting chronic diseases (i.e., arthrosis) and/or dental implantation is also increasing.

[1] National University of Science and Technology 'MISIS', Leninsky Av., 4, 119415 Moscow, Russia.
[2] I.M. Sechenov First Moscow State Medical University (Sechenov University), Bolshaya Pirogovskaya Str., 2, bld. 4, 119435 Moscow, Russia.
[3] Tohoku University, Aramaki aza Aoba 6-3, Aoba-ku, Sendai, Miyagi 980-8578, Japan.
* Corresponding author: sascha-gladkova@yandex.ru

All of these factors justify the need to search for new materials for the manufacture of endoprosthesis and other medical metal implants, and for the improvement of materials already used. The most common materials for metal bioimplants are various modifications of alloys of titanium (Ti), magnesium (Mg) and stainless steel. Such materials are subject to the highest demands, for both their strength characteristics and surface properties. The product must be biocompatible, and in some cases, it must have a sufficiently branched porous surface, and/or in other cases it should biodegrade at a given speed.

To solve these and other problems, coatings are applied to the surface(s) of products in various ways, which allows one to significantly modify the properties of the material.

The superelasticity and shape memory of some Ti alloys make it possible to create fundamentally new medical devices (i.e., stents and conductors), which gave new impetus to the development of an entire medical industry—interventional radiology. It has become possible to bring "folded form" stents into an area of vascular stenosis and straighten them to the desired diameter in patients, thus allowing both medical professionals and patients to avoid incisions and large surgical interventions. The issue of optimizing the surface properties of products based on titanium nickelide (TiNi) is also highly relevant. In the absence of a special coating on the surface of the product, nickel (Ni), which is toxic to the body, is released from the alloy into biological tissues [4].

The same effect in biological media is observed for alloys based on iron (Fe), cobalt (Co), and chromium (Cr) [5, 6]. Therefore, various surface modification technologies are offered for applying coatings with desired properties to the surface(s) of medical devices. Hydroxyapatite is a biologically active agent that has a chemical composition, structure, and physical characteristics similar to natural bone. Theoretically, its deposition on the metal surface should lead to a decrease in corrosion, an increase in biocompatibility, and an increase in the strength characteristics of the implant. However, several studies have shown that the coating of pure hydroxyapatite on the surface of stainless steel (316L SS) does not improve the corrosion resistance [7]. In contrast, the formation of a combined hydroxyapatite-based coating with the addition of graphene [8] or calcium silicate [9], improves the corrosion resistance of the substrate material [10, 11].

Obtaining a stable layer of oxides on the passivated metal surface can play a significant role in preventing further corrosion, as well as preventing and/or reducing the release of toxic metal ions from the underlying layers. Oxide film can be modified via various options of exposure to the metal surface, thickness, structure and chemical composition [12, 13]. The formation of an oxide film with desired properties, in addition to preventing corrosion, can play a significant role in ensuring the biocompatibility of the material [14].

Passivation of the metal surface can be carried out thermally, electrochemically, or in solutions of oxidizing agents (i.e., nitric acid) [15, 16, 17]. Various modifications of titanium oxide (TiO_2) are obtained on the surface of Ti alloys [18, 19, 20]. On the surface of cobalt-chromium alloys, chromium oxide (Cr_2O_3) [21, 22, 23] is obtained, and on stainless steel (316L SS), iron oxide (Fe_2O_3) [24, 25].

A passivating oxide layer can serve as the basis for applying various coatings. A number of studies have shown that applying hydroxyapatite to a passivated surface increases corrosion resistance and improves biocompatibility for stainless steel (316L SS) [26] and Ti [27, 28, 29]. If the oxide film is damaged, the outer surface can be repassivated. The time of repassivation or regeneration of the oxide film depends on the material [30, 31]. A number of studies have shown that ion-beam surface modification of metal bioimplants reduces the release of toxic ions, increases corrosion resistance and surface hardness, and improves biocompatibility [32–39]. Micro-texturing is a relatively new method for modifying the surface of bioimplants; it provides an increase in corrosion resistance, improves biocompatibility and osteointegration [40–43].

Among the newer surface modification techniques, PEO is the most promising method for multifunctional coatings formation on the light structural alloys; they are widely used in different sectors of industry to reduce the weight of products due to their relative low densities. This method allows obtaining dense, well-adhered, ceramic-like multifunctional coatings with such properties as high wear resistance, hardness, and corrosion protection. Depending on the electrolyte composition and process mode, these methods produce coatings with good biocompatibility and controlled surface porosity. Furthermore, PEO allows for the inclusion of separate elements from the composition of the electrolyte into the composition of the formed coating, of which the properties of the base material can vary significantly.

The principles of targeted formation of bioinert, bioactive and bioresorbable coatings are directed towards research in the area of material science for the creation of new implants. This chapter provides an overview about the PEO treatment, properties of multifunctional coatings and their applications in the biomedical sphere, as well as their corrosion protection.

2. Plasma Electrolytic Oxidation: An Overview

It is well known that lightweight metals such as aluminum, titanium and magnesium cannot be used without protective coatings in various aggressive medias. Materials' manufacturers have been using a surface treatment technology known as anodizing to prevent damages. By immersing metals in a solution that conducts electricity and applying an electric current, the natural oxide on the metal's surface thickens and creates a protective outer layer.

Although anodized layers are typically porous [44, 45], this simple process can alter the surface appearance of a metal and, more importantly, improve its resistance to corrosion—mostly after impregnation. For a long time, it was believed that anodic sparks are a negative phenomenon, resulting in the formation of less homogeneous and more porous films. However, in 1956–1965 Neil and Gruss [46–51] were able to use anodic sparks to synthesize complex oxide coatings based on the mixture of substrate and electrolyte components. More recent modifications have improved this process, by using higher voltage and stronger electric fields (Fig. 1).

Many researchers [52] offer various definitions/methods of plasma electrolytic oxidation (PEO). Generally, PEO is described as a novel plasma-assisted surface engineering technology, which allows growing nanostructured oxide-based coatings,

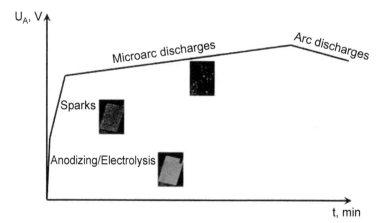

Fig. 1. Stages of the PEO process, adapted from reference [52].

by the application of an electric field greater than the dielectric breakdown field for the oxide. Short-lived micro-discharges occur, and the resulting electro-chemical and plasma-thermochemical reactions contribute to the growth of the coating. PEO may also be referred to as microarc oxidation (MAO) [52, 53] or spark (discharge) anodizing [54].

2.1 Advantages of PEO

- The possibility of obtaining coatings with a high breakdown voltage (up to 3000 V for coatings formed on Al and Al-based alloys), and adhesion to the metal substrate (up to 60 MPa);

- A significant reduction in the required production time/area—the products do not require careful surface preparation (washing, etching, degreasing, bleaching, etc., which are necessarily present in other methods, in particular, anodizing), due to the high temperature in plasma micro-discharges [42, 55];

- An eco-friendly process: the electrolytes are environmentally friendly and are free from any heavy metals [56, 57];

- Avoidance of expensive equipment required for competing vacuum-based plasma technologies;

- There is no heating of the metal substrate of the products;

- Obtaining coatings with uniform thickness and properties, on products of any geometric form and/or dimensions.

2.2 Areas of Application

This process is not only one-step and non-complex (yet extremely effective) technology for producing very thick coatings (up to 350 μm) on the substrate materials during relatively short time, it also vastly enhances their hardness, thermal resistance, dielectric strength, friction coefficient, corrosion resistance, and improves

Fig. 2. Pictures of the coated industrial details by PEO: oil-gas industry-lock of a ball stopcock (a); rail industry (b); automotive industry-parts of the internal combustion engine (c) and parts of an ABS system (f); medical industry—body of medical equipment (d) and dental implants (e).

biocompatibility and prevents wear. Today, items coated by PEO metals are used in many real-world industries and products: machinery, wind power, shipbuilding, aviation, medical, automotive, agricultural, construction, the railway sector, and consumer goods (Fig. 2).

2.3 Set-up for PEO

Typically, the process consists of applying a high voltage difference of over 250 V between a metallic 'working' electrode (detail used as anode) and a counter-electrode mostly made of stainless steel (for instance, an electrolytic container itself or a stainless-steel jacket, placed adjacent to the container wall, may be used as cathode), that are both immersed into electrolyte (Fig. 3). AC, DC or pulsed AC/DC modes with the varied ratio of anodic and cathodic parts are utilized [43, 58] (Commercial setup for PEO is shown in Fig. 3).

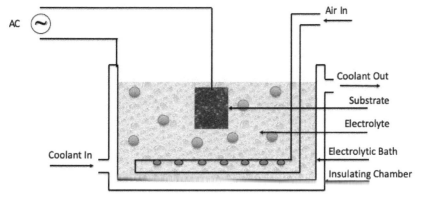

Fig. 3. A schematic diagram of a commercial setup for PEO.

196 Corrosion Protection Using Graphene and Biopolymers

There are different modifications of the setups for PEO, though most consist of a power electrical source and pulse generator, transformer; stainless steel tank/container with the electrolyte cooled with the industrial chiller or with simple laboratory water-cooling system; signal acquisition system (incl. ammeter, voltmeter) [42].

2.4 Process Description

The process of PEO consists of several stages (Fig. 1):

I. Anodizing without visible plasma discharges and/or electrolysis (simultaneous flow and electrophoresis are possible when particles, such as oxides, are introduced into the electrolyte);

II. Plasma electrolytic oxidation in the presence of spark discharges;

III. Plasma electrolytic oxidation in the presence of plasma micro-discharges on the surface of the working electrode;

IV. Plasma electrolytic oxidation in the presence of plasma arc discharges.

During the first stage, porous film is forming; despite the fact that the film is highly porous, significant increases in the anode voltage on the working electrode occurs (Fig. 1). Stages II and III differ from each other in the geometric dimensions of plasma micro-discharges, and in the intensity of ignition (Fig. 1). The mechanisms of ignition of these micro-discharges and the growth of coatings in stages II, III are identical. The temperature of micro-discharge is more than 2500°C during Stage III [59–61].

A 'separate' plasma micro-discharge (visible to the naked eye) consists of cascades of individual micro-discharges, with typical time periods of realization of about 0.1–1 ms. The maximum current in the discharge is approximately 10 to 100 mA. A key to the development of such coatings lies in the understanding and characterization of the plasma processes involved in the individual discharges. The transition of the process to stage IV 'arc discharges' is possible. At this stage, the ignition mechanism of the discharges is also identical to the mechanism of stage II and III. Due to higher energy in the discharge channels, almost all the plasma is released into the electrolyte, rather than being carried to the surface of the coating, thus leading to the formation of local, crater-like macro-defects in the coating.

The plasma chemistry of the surface discharges is complex in nature; it involves charge transfer at the substrate/electrolyte border, strong ionization and charge transfer effects between the substrate surface and the electrolyte through the oxide layer with the aid of the plasma (which plays important roles in the local melting, melt-flow, and oxidation of the substrate). A comparison between coating formation mechanisms on Al, Ti and Mg and their alloys is clearly demonstrated by Rakoch et al. and Hussein et al. [55, 61–63].

3. Wear-Resistant and Anticorrosive PEO

Due to the phase composition of the coating, as well as their microstructure, and properties are greatly affected by the nature of the alloying elements of the alloys [58, 64–66], electrical mode of PEO [42, 61, 67, 68], and electrolyte composition.

For instance, while conventional anodizing can only produce amorphous aluminum oxide (due to main differences in the applied current and voltage ranges), the PEO process can produce a material more than five times harder (1600–2000 $HV_{0.1}$). For example, the coatings on 2024 aluminum alloys is harder than any steel—or even glass and sand—making it scratch or erosion resistant.

The electrolytes for PEO, which can be easily revitalized, typically have aqueous base solutions, with a low concentration of salts and/or phosphate, silicate, aluminate and borate additives [57, 69]. It is also common to combine two or more techniques for surface coating and/or modification, to obtain properties that are unattainable using an individual technique. The data presented in Table 1 demonstrate a difference in coating growth rate, thickness, and structure, obtained by PEO in dependence with technological regimes.

Table 1. Technological regimes of PEO, properties and the most frequent average growth rates of coatings synthesized in aqueous electrolytes on the aluminum alloys.

Property	Wear-resistant anti-corrosion PEO coatings	Anti-corrosion PEO coatings
Current density, A/dm^2	10–30	4
Current mode/Frequency, Hz	AC/50	AC/50
Voltage, V	750	470
Treatment time, min	60–240	10–40
Silicate concentration, g/l	up to 20	up to 200
KOH concentration, g/l	up to 4	–
Coating structure	crystalline	amorphous
Coating growth rate, μm/min	up to 1.5	up to 4
Coating thickness, μm	up to 350	up to 80
Current density, A/dm^2	10–30	4

Hitherto, research interests have been directed towards:

- Improving mechanical and functional properties of the coatings, with the addition of nanoparticles (in particular, cerium oxide nanoparticles, silicone dioxide, diamond powder, MoS_2 particles [70–75], graphite [76], and graphene [77]), because nanoparticles included in PEO coatings may improve corrosion resistance, remarkably decrease micropores and microcracks [77];
- Photocatalytic activity [78–81] and photoluminescence properties [82, 83] of the PEO coating;
- Producing of self-lubricating PEO coatings for improving reliability of vehicles' engines [84];
- The energy efficiency of the process [85].

4. Biocompatible PEO Coatings

4.1 Bioactive PEO Coatings

PEO is a surface treatment that provides well-controlled, differing thicknesses of oxide layers with nanoporous pits, and incorporates calcium (Ca) and phosphorus (P) into the coatings on Ti alloys. The modification of the Ti alloys' surface by PEO could potentially improve the osteointegration of medical implants. In this study, [86] the *in vitro* ability of cells to adhere to PEO treated Ti alloy surfaces was investigated. Three different types of surface modifications were examined and compared: PEO, machined, and grit-blasted. To assess the structure and morphology of the surfaces, energy-dispersive X-ray spectroscopy (EDX) and scanned electron microscopy (SEM) analyses were performed. Biological and morphological responses of the osteoblast cell lines (SaOS-2) were examined by measuring the cell proliferation and differentiation (alkaline phosphatase (ALP)) activity, and $\alpha v \beta 3$ integrin level. The cell proliferation rate, ALP activity, and cell adhesion to the surface increased in the PEO group, in comparison to those in the groups of machined and grit-blasted surfaces. In addition, the osteoblast cell lines of the PEO group that were homogeneously spread across the surface were strongly adhered, and well-differentiated when compared to that of the other groups. Thus, this method could be a realistic option for treating the surfaces of Ti alloy, resulting in better osteointegration.

The same results were achieved in another investigation [57]. Previous research has shown that PEO treatments provide a promising approach to improving the integration of Ti implants with their surrounding tissues. In this study, PEO-coating was applied to the surface of Ti implants in order to examine the resulting microstructure, as well as to characterize the adhesion and viability of cells on the treated surfaces. The main goal of this study was to identify an optimal surface modification for Ti implants, in order to improve soft-tissue integration. Three variants of coating were generated on Ti alloy Ti6Al4V via PEO. The elemental composition and the microstructures of the surfaces were characterized using EDX, SEM and profilometry. Cytocompatibility of the coatings *in vitro* was assessed by seeding fibroblasts (L929) onto the investigated surface, and measuring the adhesion, viability and cytotoxicity of cells by means of live/dead staining, XTT assay and LDH assay. Electron microscopy and profilometry revealed that the PEO-surface variants differed largely from the untreated control material as well as from one another, in microstructure/topography, porosity and roughness. Roughness generally increased after PEO-treatment. *In vitro*, PEO-treatment led to improved cellular adhesion and viability of cells, accompanied by decreased cytotoxicity.

In another previous study [87], the surface characteristics and electrochemical properties of bioactive coatings produced by PEO in solutions with Ca, P, silicon (Si) and silver (Ag) on commercially pure Ti were investigated. The results showed a porous oxide layer, produced using PEO treatment, improving the surface topography. PEO treatment enriches the surface with bioactive elements, responsible for mimicking bone surface; the surfaces with higher calcium concentrations presented antibacterial and biocompatibility properties with better responses for corrosion and barrier properties, due to the presence of rutile crystalline structure.

Previous research determined PEO could be a promising surface treatment option to improve the electrochemical behavior of implanted medical Ti devices, for example for dental implants mitigating treatment failures.

Besides the enrichment of coating with the bioactive elements, physical structure modification of the surface could be one of the possibilities of improving of the Ti medical implants biocompatibility. The sub-microporous PEO coating on Ti implant surface with micro-scale gouges has been formed using a PEO process, in order to overcome the compromised bone-implant integration. The treated implant has been further mediated by post-heat modification to compare the effects of -OH functional group and the nano-scale orange peel-like morphology on osteointegration. The results showed the greatly improved push-out force for the PEO coated Ti implants with micro-scale gouges. That could be attributed to the excellent mechanical interlocking effect between implants and biologically meshed bone tissues. OH functional group promoted synostosis between the biologically meshed bone and the gouge surface of implant [88].

The effects of the PEO surface on seeded cell morphology and cytoskeleton were investigated in previous work [89]. Four groups of pure Ti samples were prepared: grooved surfaces (group G); sandblasted surfaces (group SB); grooved PEO surfaces (group GPEO); and sandblasted-PEO-surfaces (group SBPEO). Each sample had a diameter of 15 mm and a thickness of 1 mm. Osteoblast cells were cultured on the surface of each sample. The morphology and proliferation of the cells on the Ti surfaces were observed via SEM. The cytoskeleton was evaluated using a laser scanning confocal microscope (LSCM). The osteoblasts were inoculated after 12 h, then spread along the surface of the Ti plate. In groups GPEO and SBPEO, the osteoblasts converged in the holes. The actin fibers in each group were clearly visible. In particular, the actin fibers in the GPEO and SBPEO groups were arranged in parallel and formed bundles that extended into the holes. Thus, these results show PEO Ti surfaces significantly affected the morphology and cytoskeleton of seeded osteoblasts.

3D-printed porous Ti holds enormous potential for load-bearing orthopedic applications. Although the 3D printing technique has respectable control over the macro-structures of porous Ti, the surface properties that affect tissue response are beyond its control, thus adding the need for tailored surface treatment to improve its osteointegration capacity. PEO exhibits a high efficiency in the enhancement of osteointegration of porous Ti64, via optimizing the patterns of bone ingrowth and bone/implant interlocking. Therefore, post-treatment of 3D printed porous Ti64 with PEO technology could generate several possibilities for the development of bioactive customized implants in orthopedic applications. Previous studies [90] have applied the one-step PEO process to a 3D-printed porous Ti6Al4V scaffold, in order to endow the scaffold with a homogeneous layer of microporous TiO_2, in addition to significant amounts of amorphous calcium-phosphate. Following the treatments, the porous Ti64 scaffolds exhibited a drastically improved apatite forming ability, cytocompatibility, and alkaline phosphatase activity. For example, *in vivo* testing in a rabbit model showed that the bone ingrowth at the untreated scaffold was in a pattern of distance osteogenesis, by which bone formed only at the periphery of the scaffold. In contrast, the bone in-growth at the PEO-treated scaffold exhibited a pattern of contact osteogenesis, by which bone formed *in situ* on the entire surface of the scaffold. This pattern of bone in-growth significantly increased bone formation both in and around the scaffold, possibly through enhancement of bone formation and disruption of bone remodeling. Moreover, the implant surface of the PEO-treated scaffold interlocked with the bone tissues through the fabricated microporous topographies to generate a stronger bone/implant interface. The results of the push out test showed the increased osteo integration strength.

In previous research by Li et al. [91], PEO coating on Ti was performed as a scaffold for hydroxyapatite layer. A thin hydroxyapatite (HA) layer was coated on a PEO treated Ti (PEO-Ti) substrate by means of the sol-gel method. The PEO enhanced the biocompatibility of the Ti, and the bioactivity was improved further by the HA sol-gel coating. The HA sol was aged fully, to obtain a stable and phase-pure HA; the sol concentration was varied to alter the coating thickness. As a result of the HA sol-gel coating, the Ca and P concentrations in the coating layer increased significantly. However, the porous morphology and roughness of the PEO-Ti was altered very little by the HA sol-gel treatment. The proliferation and alkaline phosphatase (ALP) activity of the osteoblast-like cells on the PEO/HA sol-gel-treated Ti was significantly higher than those on the PEO-Ti without the HA sol-gel treatment.

Another possibility for further improvements of the biocompatibility of PEO coatings on Ti is through electrochemically reducing in an alkaline solution, proposed in G. Li's previous work [92]. During the first stage, the oxide layer on the Ti surface was formed by PEO, then the coating was electrochemically reduced in an alkaline solution (PEO-AK). A series of *in vitro* stem cell differentiation, and *in vivo* osteointegration experiments, were carried out to evaluate the osteogenic capacity of the resulting coatings. *In vitro*, the initial adhesion of the canine bone marrow stem cells (BMSCs) seeded on the PEO and PEO-AK coatings was significantly enhanced, and cell proliferation was promoted. In addition, the expression levels of osteogenesis-related genes, osteorix, alkaline phosphates (ALP), osteopontin, and osteocalcin, in the canine BMSCs, were all up-regulated after incubation on these coatings, especially on the PEO-AK coating. In addition, the *in vitro* ALP activity and mineralization capacity of canine BMSC cultured on the PEO-AK group was better than that on the PEO group. Furthermore, six weeks after the insertion of the Ti implants into canine femurs, both the bone formation speed and the bone-implant contact ratio of the PEO-AK group were significantly higher than those of the PEO group. Thus, all of these results suggest that this duplex coating process is promising for engineering Ti surfaces to promote osteointegration for dental and orthopedic applications.

Titanium is widely used in biomedical materials, particularly in dental implants, because of its excellent biocompatibility and mechanical characteristics. However, Ti implant failures, in some cases, vary with implantation sites and patients. Improving its overall performance is a major focus of dental implant research. Equal-channel angular pressing (ECAP) can result in ultrafine-grained Ti with superior mechanical properties and better biocompatibility, which significantly benefits dental implants, without any harmful alloying elements. Lanthanum (La) can inhibit the acidogenicity of dental plaque; La-containing hydroxyapatite (La-HA) possesses a series of attractive properties, in contrast to La-free HA. PEO allows the inclusion of elements from the electrolyte solution into the structure of coating. Authors Deng et al. [93] proposed preparation porous La-containing hydroxyapatite coatings with different La contents (0.89%, 1.3%, 1.79%) on ultrafine-grained ($\sim$ 200–400 nm) Ti, using ECAP and PEO in electrolytic solution containing 0.2 mol/L calcium acetate, 0.02 mol/L β-glycerol phosphate disodium salt pentahydrate (β-GP), and lanthanum nitrate, with different concentrations to further improve the overall performance of Ti; these are expected to show great potential in medical applications as dental implants.

It is possible to include calcium (Ca^{2+}), magnesium (Mg^{2+}) and phosphate (PO_4^{3-}), hydrogen phosphate (HPO_4^{2-}), dihydrogen phosphate ($H_2PO_4^-$), pyrophosphate ($P_2O_7^{4-}$) as well as zinc (Zn^{2+}) or copper (Cu^+/Cu^{2+}) ions in the PEO coating on Ti (described by Rokosz et al. [69]), thought that the obtained surface should be characterized by high biocompatibility, due to the presence of structures based on Ca and phosphates, and have bactericidal properties, due to the presence of zinc and copper ions. Furthermore, the addition of Mg ions should accelerate the healing of postoperative wounds, which could lead to faster patient recovery. The characteristics of porous coatings fabricated at three voltages in electrolytes based on H_3PO_4 with calcium nitrate tetrahydrate, magnesium nitrate hexahydrate, and copper (II) nitrate or zinc trihydrate, were investigated. SEM, energy dispersive spectroscopy (EDS), glow discharge optical emission spectroscopy, X-ray photoelectron spectroscopy, and X-ray diffraction (XRD) for coating identification were used. It was found that the higher the voltage during PEO, the thicker the porous coating was, with higher amounts of built-in elements coming from the electrolyte, and a more amorphous phase with signals from crystalline $Ca(H_2PO_4)_2 \cdot H_2O$ and/or $Ti(HPO_4)_2 \cdot H_2O$. Additionally, the external parts of the obtained porous coatings formed on Ti consisted mainly of Ti^{4+}, Ca^{2+}, Mg^{2+} and PO_4^{3-}, HPO_4^{2-}, $H_2PO_4^-$, $P_2O_7^{4-}$ as well as Zn^{2+} or Cu^+/Cu^{2+}.

In a study by Zhang et al. [94], the bioactivity of manganese-incorporated TiO_2 (Mn-TiO_2) coating prepared on Ti plates by PEO in Ca⁻, P⁻ and Mn⁻containing electrolytes was investigated. The study showed that the Mn-TiO_2 coating has potential for orthopedic implant applications. The topography of the surface, element and phase compositions was investigated using SEM, EDS, and XRD. The adhesion of MG63 osteoblast-like cells onto untreated Ti, TiO_2 and Mn-TiO_2 coatings was performed; the signal transduction pathway involved was confirmed by the sequential expression of the genes for integrins β_1, β_3, α_1 and α_3, focal adhesion kinase, and the extracellular regulated kinases (ERKs), including ERK1 and ERK2. The data showed that Manganese (Mn) was successfully incorporated into the porous nanostructured TiO_2 coating and did not alter the surface topography or the phase composition of the coating. The adhesion of the MG63 cells onto the Mn-incorporated TiO_2 coating was significantly enhanced, compared with that on the Mn-free TiO_2 coating and the pure Ti plates. The enhanced cell adhesion on the Mn-TiO_2 coatings may have been mediated by the binding of the integrin subunits, β_1 and α_1, and the subsequent signal transduction pathway, involving FAK and ERK?

Loosening of pedicle screws in osteoporotic bones causes failure of fixation. This is an issue in spinal surgery. Pedicle screws with a bioactive surface treated by PEO can improve screw fixation strength in bones. Shi et al. [90] compared *in vivo*, the fixation strength between pedicle screws treated with PEO and untreated screws in an osteoporotic model of sheep. The PEO treated and untreated screws were placed in lumbar vertebral bodies. After three months of implantation, biomechanical tests, micro-CT analysis, and histological observations were conducted to examine the performance of the two groups. Initially, no significant difference was found between the two groups in biomechanical tests ($p > 0.05$); three months later, higher pull-out strength and load with less displacement were detected in the PEO-treated group ($p < 0.05$). Micro-CT analysis showed that the tissue mineral density, bone volume fraction, trabecular thickness, and trabecular number in the PEO-treated group were all higher than those in the untreated group ($p < 0.05$), and trabecular spacing was smaller ($p < 0.05$). Histologically, the bone-implant interface in the PEO-treated group was better than that in untreated group ($p < 0.05$).

Aside from an improvement of the surface properties of orthopedic and dental implants, PEO could be useful for surface modification of endovascular devises such as valve prostheses, vascular stents, etc. Ultrafine-grained pure Ti prepared by equal-channel angular pressing has favorable mechanical performance and does not contain alloy elements that are toxic to the human body. Thus, it has a potential clinical value in applications such as cardiac valve prostheses, and vascular stents. To overcome the material's inherent thrombogenicity, surface-coating modification is a crucial pathway to enhance blood compatibility. The PEO method can significantly enhance the blood compatibility of ultrafine-grained pure Ti, increasing its potential for practical applications in cardiovascular surgery. An electrolyte solution of sodium silicate + sodium polyphosphate + calcium acetate and the PEO technique were employed by Xu et al. [95] for *in situ* oxidation of an ultrafine-grained pure Ti surface. A porous coating with anatase- and rutile-phase TiO_2 was generated, and wettability and blood compatibility were examined. The results showed that, in comparison with ultrafine-grained pure Ti substrate, the PEO coating had a rougher surface, smaller contact angles for distilled water, and higher surface energy. PEO modification effectively reduced the hemolysis rate, extended the dynamic coagulation time, prothrombin time (PT), and activated partial thromboplastin time (APTT), reduced the amount of platelet adhesion and the degree of deformation, and enhanced blood compatibility. In particular, the sample with an oxidation time of nine minutes possessed the highest surface energy, largest PT and APTT values, smallest hemolysis rate, less platelet adhesion, a lesser degree of deformation, and a more favorable blood compatibility.

Stainless steel is one of the most widely used biomaterials for internal fixation devices, but it is not used in cementless arthroplasty implants because a stable oxide layer (essential for biocompatibility) cannot be formed on the surface. Modification of the surface using Ti electron beam coating and PEO was performed to improve the ability of stainless-steel implants to osteointegrate. Ti electron beam coating, to form an oxide layer on the stainless-steel surface, was performed by Lim et al. [96]. To form a thicker oxide layer, a PEO process was used on the surface of Ti coated stainless steel. The ability of cells to adhere to grit-blasted, Ti-coated, PEO treated stainless steel *in vitro*, was compared with that of two different types of surface modifications: machined and Ti-coated, and PEO. EDX and SEM investigations were used to assess the chemical composition and structure of the stainless-steel surfaces and cell morphology. The biological responses of an osteoblast-like cell line (SaOS-2) were examined by measuring proliferation (cell proliferation assay), differentiation (alkaline phosphatase activity), and attraction ability (cell migration assay). Cell proliferation, alkaline phosphatase activity, migration, and adhesion were increased in the grit-blasted, Ti-coated, PEO group, compared to the two other groups. Osteoblast-like cells on the grit-blasted, Ti-coated, PEO surface were strongly adhered, and proliferated well, compared to those on the other surfaces. It was assumed that the process is not unique to stainless steel, that it can be applied to many metals to improve their biocompatibility, thus allowing a broad range of materials to be used for cementless implants.

Besides Ti, Zirconium (Zr) is a promising material for medical implants. The biocompatibility of Zr is at least on par with that of Ti. Plasma electrolytic oxidation in solutions based on calcium acetate and calcium β-glycerophosphate could improve the bioactivity of Zr surfaces. In a previous study by Sowa et al. [97], the process of direct current (DC) PEO of pure Zr in the electrolyte, based on $Ca(H_2PO_2)_2$, $Ca(HCOO)_2$, and $Mg(CH_3COO)_2$, was investigated. The process was conducted at 75 mA/cm^2 up to 200, 300, or 400 V. Five stages of the process were revealed. The process at higher voltages lead to the formation of oxide layers with Ca/P or (Mg+Ca)/P ratios close to that of hydroxyapatite (Ca/P = 1.67), which were determined by SEM and EDX. The corrosion resistance was investigated using electrochemical impedance spectroscopy (EIS) and DC polarization methods. A $R(Q[R(QR)])$ circuit model was used to fit the EIS data. The results present the improving of corrosion and pitting resistance, regardless of the applied process condition compared to pure Zr surface.

This study [40] was performed in order to gain a better understanding of the relationship between PEO discharge types, and the bioactivity of an oxide coating synthesized on a Zr scaffold. The types of discharge and the coating growth mechanism were identified by the examination of the real cross-section image of the coating microstructure. The coating was formed by using PEO in an electrolyte solution with Na_2SiO_3, $Ca(CH_3COO)_2$ and $C_3H_7Na_2O_6P$. The processing times were: 2.5, 5, 15, and 30 min. The effect of the process duration on the different discharge model types (Type-A, B, and C) and bioactivity of the coatings was investigated by using XRD, Fourier Transform Infrared Spectroscopy (FTIR), Scanning Electron Microscopy-Energy-Dispersive X-ray spectroscopy measurements (SEM-EDS) and Optical Surface Profilometry (OSP). It was shown that increasing the duration of PEO resulted in thicker and rougher coatings. The XRD data revealed that all of the samples prepared at differing process durations contained the t-ZrO$_2$ (tetragonal zirconia) phase. During the PEO process, non-crystalline hydroxyapatite formed, which was confirmed through the FTIR data. The surface morphology as well as the amount and distribution of the features of the coating surface were modified by increasing voltage. The simulated body fluid tests showed that a more bioactive surface with more HA crystals formed, due to the chemical composition and high surface roughness of the coating. The pore, crack and discharge structures played a key role in apatite nucleation and growth and provided ingrowth of apatite into discharge channels on the coatings' surface.

Attempts are being made to obtain PEO coatings on other metals to increase its biocompatibility. For example, PEO treating of tantalum surfaces is described in previous studies [98]. The development of an oxidized surface layer on tantalum, similar to those of existing oral implants on the market today, inspired this work. SEM images and EDS analysis revealed the changes on the tantalum surface according to different exposure times and confirmed increased salt deposition with the time of PEO process.

4.2 Biodegradable PEO Coatings

Pure Mg and Mg alloys with improved corrosion resistance appeal to the increasing interests of it as a revolutionary biodegradable material, for fractured bone-fixing implants. However, the *in vivo* corrosion degradation of the Mg and bone healing response in the metal-bone border are not well understood. The understanding of mechanism and dynamic of this process is greatly important for the medical application of Mg. Besides the highest biocompatibility, well-controlled, stable and projected dynamic of biodegradation of bioresorbable implants will be necessary for medical application. PEO is a process that allows controlling the properties of a coating on a metal surface; thus, it could be a promising method of controlling biodegradation dynamic of Mg implants.

Previous research [99] reports improved mechanical properties of Mg alloys (AZ31) by PEO in electrolytes with NaOH, Na_2SiO_3, KF and $NaH_2PO_4 \cdot 2H_2O$. Mechanical properties such as wear resistance, surface hardness and elastic modulus were increased for PEO-coated AZ31 Mg alloys (PEO-AZ31). DC polarization in Hank's solution indicated significant growth in the corrosion resistance for PEO-coating in KF-contained electrolyte.

Magnesium (AZ31) implants with 10 μm and 20 μm thick biocompatible PEO coatings, mainly composed of MgO, Mg_2SiO_4, $CaSiO_3$ and $Mg_3(PO_4)_2$ phases, were investigated *in vivo* in animal models [100]. The electrochemical tests showed an improved corrosion resistance of Mg as a result of the PEO coatings. The Mg implants' degradation and uninhibited bone healing were investigated using a 3 mm-wide bone fracture defect model. The uncoated AZ131 plates, and plates with a 10 μm and 20 μm PEO coating, were implanted into the site of radius bone fracture in adult New Zealand white rabbits. The blood test and histological examination did not show any negative side effects. The groups with implanted coated and uncoated AZ131 plates showed the promoting effect of bone fracture healing, compared with that of the simple fracture group without implants. The release of Mg^{2+} ions by the degradation of the implants into the fracture site improved the bone fracture healing, which is attributed to the Mg promoting CGRP-mediated osteogenic differentiation.

Biodegradation and osteogenic properties of a PEO coated AZ31 Mg alloy was evaluated *in vivo* for the fracture fixation in the animal models. The samples with different coating thicknesses applied using PEO were used for treatment of the bilateral radial fracture in thirty-six New Zealand white rabbits. The animals were randomly divided into A, B, and C groups at four points in time; another three rabbits without AZ31 implants served as the control group. Coated Mg alloy AZ31 samples were implanted on the fracture and stitched with silk thread. Data was collected for general observation, histology, X-ray, hematology, and mechanical properties; changes were detected at the second, fourth, eighth, and twelfth week after implantation.

The fractures in each rabbit were healed by the twelfth week after implantation. The best results for general observation, histology, and X-ray appeared in the A group (Mg alloys implants without coating). However, the A group showed the worst results from the perspective of mechanical properties; tensile strength and flexural strength failed to reach that of the natural bone by the twelfth week. Comprehensive results displayed that the C group (20-μm PEO) coating results were better than the others for mechanical properties, and that there was no difference between the B and C groups in hematology. The biodegradation rate of PEO coated AZ31 *in vivo* was inversely proportional to the coating thickness.

The influence on surface coatings of two electrolyte solution additives, hexamethylenetetramine and mannitol, was investigated in the study [101]. The results of *in vitro* studies in NaCl showed an improvement in the corrosion resistance.

Effects of the alloying element Ca, on the bioactivity and corrosion properties of coatings formed by PEO on Mg AM60 alloys, were investigated [102]. The corrosion resistance was studied by conducting electrochemical tests in 0.9% NaCl solution. The biodegradation process was evaluated by soaking the samples in simulated body fluid. Under identical conditions, the PEO coating thicknesses increased, with an increasing Ca content in the alloys. Thicker apatite layers grew on the PEO films of Ca-containing alloys, which improved the corrosion resistance in the NaCl solution.

The PEO coating with a MgF_2 phase, could be a promising surface modification of Mg alloys for medical application, if the content of fluoride will be well controlled. The corrosion and biological properties of fluoride-incorporated PEO coating on a biodegradable AZ31 alloy were described [103]. The surface structure of the coating was investigated by SEM and XRD. The corrosion behavior was investigated in simulated body fluid, and *in vitro* cytocompatibility of the coatings was also studied by evaluating cytotoxicity, adhesion, proliferation and live–dead stain of MC3T3-E1osteoblast cells. The corrosion properties *in vivo* were also examined. The results showed that fluoride persisted in the formed coating in the structure, during the MgF_2 phase. *In vitro* and *in vivo* biodegradation tests showed that the corrosion resistance of the coatings increased with the incorporation of fluoride; this is most likely due to the chemical stability of the MgF_2 phase. Thus, the good cytocompatibility of fluoride-incorporated coatings was confirmed. However, when the fluoride content was high, a slight inhibition of cell growth was observed. The results indicate that although fluoride incorporation can enhance the corrosion resistance of the coatings (making a more suitable local condition for surrounding cells), a high content of fluoride in the coating could kill the surrounding cells, due to the high release rate of fluorine.

Mg ions could be incorporated in the coating structure from the electrolyte solution. The influence of electrolyte solution additives (magnesium nitrate hexahydrate ($Mg(NO_3)_2 \cdot 6H_2O$) and/or zinc nitrate hexahydrate ($Zn(NO_3)_2 \cdot 6H_2O$) in concentrated phosphoric acid (H_3PO_4 (85% w/w))) on the PEO coatings' structure and properties was investigated [104]. Complementary methods were used for the formatted coatings' surface characterization: SEM, EDX, X-ray photoelectron spectroscopy (XPS), glow discharge optical emission spectroscopy (GDOES) and XRD. The results showed that increasing the contents of the additives (250 g/L $Mg(NO_3)_2 \cdot 6H_2O$ and 250 g/L $Zn(NO_3)_2 \cdot 6H_2O$) in the electrolyte solution lead to increasing Mg/P and Zn/P ratios, as well as increasing the coating thickness. In addition, it was found that by increasing the PEO voltage, the Zn/P and Mg/P ratios increase as well. The analysis of XPS spectra showed the existence of Mg^{2+}, Zn^{2+}, Ti^{4+}, and phosphorus compounds (PO_4^{3-}, or HPO_4^{2-}, or $H_2PO_4^-$, or $P_2O_7^{4-}$) in 10 nm top of coating.

Aside from the ions of inorganic compounds, the PEO process could be a promising means for including organic molecules in the coating structure, which, potentially, could be necessary in the future creation of biodegradable materials. The properties of PEO coatings obtained via the Mg alloy (Mg-1Li-1Ca) PEO treatment in phytic acid (PA) and poly (L-lactic acid) (PLLA) were investigated by the authors [105]. The coating surfaces were studied through field emission scanning electron microscopy (FE-SEM), electron probe X-ray microanalysis (EPMA), EDS and XRD. The corrosion resistances of the samples were evaluated via hydrogen evolution, potentiodynamic polarization and electrochemical impedance spectroscopy (EIS) in Hanks' solution. It was evident that the PEO/PLLA composite coatings significantly improved the corrosion resistance of the alloy Mg-1Li-1Ca. MTT and ALP assays, using cell culture (MC3T3 osteoblasts), showed that the PEO/PLLA coatings greatly improved the cytocompatibility *in vitro*; the morphology of the cells cultured on different samples exhibited good adhesion as well. In addition, a low hemolysis ratio in the hemolysis tests has been displayed.

Potential improvement of the PEO coating anticorrosion properties could be through hydrothermal treatments. In previous works [106], the pores of PEO coatings on AZ31 Mg alloys with Mg-Al layered double hydroxide (LDH) were successfully sealed by the hydrothermal treatment. PEO/LDH composite coating possesses a two-layer structure: an inner layer made up of a PEO coating ($\sim$ 5 μm in thickness), and an outer layer of Mg-Al LDH ($\sim$ 2 μm in thickness). Electrochemical and hydrogen evolution tests showed better corrosion resistance of the PEO/LDH coating. Cytotoxicity, cell adhesion, live/dead staining and proliferation data of rat bone marrow stem cells demonstrated that PEO/LDH coating remarkably improved the cytocompatibility of the alloy. Hemolysis rate (HR) tests show that the HR value of the PEO/LDH coating is 1.10 ± 0.47%, thus fulfilling the requirements of clinical application. The structure of Mg-Al LDH on the top of the PEO coatings showed excellent drug delivery ability as well.

Another method to potentially improve the surfaces of inorganic PEO coatings is by growing organic coatings on its surface; this has been investigated based on microstructural interpretation, electrochemical assessment, and quantum chemical analysis [107]. For this purpose, inorganic coatings with magnesium aluminate, magnesium oxide, and titanium dioxide were prepared on a Mg alloy by PEO; then, a subsequent dip-coating method was used to tailor organic coatings with diethyl-5-hydroxyisophthalate (DEIP) as organic molecules. The incorporation of TiO_2 particles worked as a sealing agent to block the micro-defects, which resulted mainly from the intense plasma sparks during PEO. Such incorporation also played an important role in increasing the adhesion between inorganic and organic coatings. The use of DEIP as an organic corrosion inhibitor resulted in a significant decrease in the porosity of the inorganic coating. Quantum chemical calculation was used to clarify the corrosion inhibition mechanism, which was activated by introduction of DEIP. Thus, the electrochemical analysis (based on potentiodynamic polarization and impedance spectroscopy tests in 3.5 wt% NaCl solution) suggested that corrosion resistance of Mg alloy samples was enhanced significantly, due to a synergistic effect arising from the hybrid inorganic and organic coatings. This phenomenon was explained in relation to electron transfer behavior, between inorganic and organic coatings.

To date, high interest of researchers is focused on investigations of the surface antibacterial activity of the coating. The increase in the number of antibiotic-resistant infections nowadays is a global problem. Modifying the surface of orthopedic implants (i.e., endoprosthesis, plates, screws, etc.) by applying nanoparticle coatings is possible solution for preventing and/or avoiding postoperative infections [108]. The antimicrobial properties of various nanoparticles are known and determined by their size, shape, charge, concentration, and reactive oxygen species generation. Nonspecific cytotoxicity of nanoparticles is distributed to both the infectious agents and to the cells of body tissues, thus imposing restrictions on the possibilities of their use in medicine [109, 110].

To modify the surface of orthopedic products in order to impart antibacterial properties to them, nanoparticles of silver (Ag) [111], gold (Au) [112], Cu [113], Zr [114], Fe [115], TiO_2 [116], zinc oxide (ZnO) [117] can be introduced into PEO coatings. PEO might be the most promising technology for incorporation of these particles into the structure of the coating. The key problem exists with the increase in the cytotoxicity of the nanoparticles, and, consequently, in its antimicrobial activity; the toxic effect on the cells of the body increases, as the biocompatibility decreases. In order to determine optimal solutions for preventing antibiotic-resistant infections, more specifically, using orthopedic medical devices based on nano-coatings, further *in vivo* and clinical trials are needed.

5. Conclusions

This chapter provides an overview on the PEO and summarizes published research data about PEO coating application, as well as ways for improving the properties of coating (i.e., wear resistance, corrosion resistance, biocompatibility, biodegradation dynamic, etc.). From the data, it can be concluded that PEO coatings are a very versatile technology, which presents a wide range of solutions for the modification of light construction metals, for a multitude of different applications. In addition, PEO is an economically viable, eco-friendly method for forming multifunctional coatings with different properties.

References

[1] Curran, A. and T. W. Clyne. 2005. Thermo-physical properties of plasma electrolytic oxide coatings on aluminium. Coat. Technol. 199: 168–179.
[2] Park, R. M., J. F. Bena, L. T. Stayner, R. J. Smith, H. J. Gibb and P. S. J. Lees. 2004. Hexavalent chromium and lung cancer in the chromate industry, a quantitative risk assessment. Risk Anal. 24: 1099–1108.
[3] Walsh, F. C., C. T. J. Low, R. J. K. Wood, K. T. Stevens, J. Archer, A. R. Poeton et al. 2009. Plasma electrolytic oxidation (PEO) for production of anodised coatings on lightweight metal (Al, Mg, Ti) alloys. Trans. Inst. Met. Finish. 87: 122–35.
[4] Hussain, H. D., S. D. Ajith and P. Goel. 2016. Nickel release from stainless steel and nickel titanium archwires—an *in vitro* study. J. Oral Biol. Craniofac. Res. 6: 213–218.
[5] Espallargas, N., C. Torres and A. I. Muñoz. 2015. A metal ion release study of CoCrMo exposed to corrosion and tribocorrosion conditions in simulated body fluids. Wear. 332-333: 669–678.
[6] Delaunay, C., I. Petit, I. Learmonth, P. Oger and P. Vendittoli. 2010. Metal-on-metal bearings total hip arthroplasty: the cobalt and chromium ions release concern. Orthop. Traumatol. Surg. Res. 96: 894–904.
[7] Robin, A., G. Silva and J. L. Rosa. 2013. Corrosion behavior of HA-316L SS biocomposites in aqueous solutions. Mater. Res. 16: 1254–1259.
[8] Janković, A., S. Eraković, M. Vukašinović-Sekulić, V. Mišković-Stanković, S. J. Park and K. Y. Rhee. 2015. Graphene-based antibacterial composite coatings electrodeposited on titanium for biomedical applications. Prog. Org. Coat. 83: 1–10.
[9] Huang, Y., S. Han, X. Pang, Q. Ding and Y. Yan. 2013. Electrodeposition of porous hydroxyapatite/calcium silicate composite coating on titanium for biomedical applications. Appl. Surf. Sci. 271: 299–302.
[10] Sridhar, T. M., U. K. Mudali and M. Subbaiyan. 2013. Preparation and characterisation of electrophoretically deposited hydroxyapatite coatings on type 316L stainless steel. Corros. Sci. 45: 237–252.
[11] García, C., S. Ceré and A. Durán. 2004. Bioactive coatings prepared by sol–gel on stainless steel 316L. J. Non-Cryst. Solids 348: 218–224.
[12] Shih, C. C., C. M. Shih, Y. Y. Su, L. H. J. Su, M. S. Chang and S. J. Lin. 2004. Effect of surface oxide properties on corrosion resistance of 316L stainless steel for biomedical applications. Corros. Sci. 46: 427–441.
[13] Trepanier, C., M. Tabrizian, L. H. Yahia, L. Bilodeau and D. L. Piron. 1999. Effect of modification of oxide layer on NiTi stent corrosion resistance. J. Biomed. Mater. Res. 48: 96–98.
[14] Ryan, G., A. Pandit and D. P. Apatsidis. 2006. Fabrication methods of porous metals for use in orthopaedic applications. Biomat. 27: 2651–2670.
[15] Simka, W., M. Kaczmarek, A. Baron-Wiecheć, G. Nawrat, J. Marciniak and J. Żak. 2010. Electropolishing and passivation of NiTi shape memory alloy. Electrochim. Acta 55: 2437–244.
[16] Huang, C. H., J. J. Lai, J. C. Huang, C. H. Lin and J. S. C. Jang. 2016. Effects of Cu content on electrochemical response in Ti-based metallic glasses under simulated body fluid. Mater. Sci. Eng. C 62: 368–376.
[17] Huang, C. H., J. J. Lai, T. Y. Wei, Y. H. Chen, X. Wang, S. Y. Kuan et al. 2015. Improvement of bio-corrosion resistance for Ti42Zr40Si15Ta3 metallic glasses in simulated body fluid by annealing within supercooled liquid region. Mater. Sci. Eng. C 52: 144–150.
[18] Jamesh, M., T. S. Narayanan and P. K. Chu. 2013. Thermal oxidation of titanium: evaluation of corrosion resistance as a function of cooling rate. Mater. Chem. Phys. 138: 565–572.
[19] Jamesh, M., S. Kumar and T. S. Narayanan. 2012. Effect of thermal oxidation on corrosion resistance of commercially pure titanium in acid medium. J. Mater. Eng. Perform. 21: 900–906.

[20] Kumar, S., T. S. N. S. Narayanan, S. G. S. Raman and S. K. Seshadri. 2009. Thermal oxidation of CP-Ti: evaluation of characteristics and corrosion resistance as a function of treatment time. Mater. Sci. Eng. C. 29: 1942–1949.

[21] Milošev, I. and H. H. Strehblow. 2003. The composition of the surface passive film formed on CoCrMo alloy in simulated physiological solution. Electrochim. Acta 48: 2767–2774.

[22] Izman, S., M. Hassan, M. R. A. Kadir, M. Abdullah, M. Anwar, A. Shah et al. 2012. Effect of pretreatment process on thermal oxidation of biomedical grade cobalt based alloy. Adv. Mater. Res. 399-401: 1564–1567. https://doi.org/10.4028/www.scientific.net/MSF.916.170.

[23] Bettini, E., C. Leygraf and J. Pan. 2013. Nature of current increase for a CoCrMo alloy: "transpassive" dissolution vs. water oxidation. Int. J. Electrochem. Sci. 8: 11791–11804.

[24] Guillamet, R., J. Lopitaux, B. Hannoyer and M. Lenglet. 1993. Oxidation of stainless steels (AISI 304 and 316) at high temperature. Influence on the metallic substratum. Le Journal de Physique IV 3 C. 9: 349–356.

[25] Souier, T., F. Martin, C. Bataillon and J. Cousty. 2010. Local electrical characteristics of passive films formed on stainless steel surfaces by current sensing atomic force microscopy. Appl. Surf. Sci. 256: 2434–2439.

[26] Gopi, D., V. C. A. Prakash and L. Kavitha. 2009. Evaluation of hydroxyapatite coatings on borate passivated 316L SS in Ringer's solution. Mater. Sci. Eng. C. 29: 955–958.

[27] Albayrak, O., O. El-Atwani and S. Altintas. 2008. Hydroxyapatite coating on titanium substrate by electrophoretic deposition method: effects of titanium dioxide inner layer on adhesion strength and hydroxyapatite decomposition. Surf. Coat. Technol. 202: 2482–2487.

[28] Narayanan, T. S., I. S. Park and M. H. Lee. 2014. Strategies to improve the corrosion resistance of microarc oxidation (MAO) coated magnesium alloys for degradable implants: prospects and challenges. Prog. Mater. Sci. 60: 1–71.

[29] Sridhar, T., S. Vinodhini, U. K. Mudali, B. Venkatachalapathy and K. Ravichandran. 2016. Load-bearing metallic implants: electrochemical characterisation of corrosion phenomena. Mater. Technol. 31: 1–14.

[30] Hanawa, T. 1999. *In vivo* metallic biomaterials and surface modification. Mater. Sci. Eng. A 267: 260–266.

[31] Hanawa, T., K. Asami and K. Asaoka. 1998. Repassivation of titanium and surface oxide film regenerated in simulated bioliquid. J. Biomed. Mater. Res. 40: 530–538.

[32] Chu, P. K., J. Chen, L. Wang and N. Huang. 2002. Plasma-surface modification of biomaterials. Mater. Sci. Eng. R. Rep. 36: 143–206.

[33] Zhao, Y., S. M. Wong, H. M. Wong, S. Wu, T. Hu, K. W. Yeung et al. 2013. Effects of carbon and nitrogen plasma immersion ion implantation on *in vitro* and *in vivo* biocompatibility of titanium alloy. ACS Appl. Mater. Interfaces 5: 1510–1516.

[34] Cui, F. and Z. Luo. 1999. Biomaterials modification by ion-beam processing. Surf. Coat. Technol. 112: 278–285.

[35] Tan, L., R. Dodd and W. Crone. 2003. Corrosion and wear-corrosion behavior of NiTi modified by plasma source ion implantation. Biomat. 24: 3931–3939.

[36] Lanning, B. R. and R. Wei. 2004. High intensity plasma ion nitriding of orthopedic materials: part II. Microstructural analysis. Surf. Coat. Technol. 186: 314–319.

[37] Wei, R., T. Booker, C. Rincon and J. Arps. 2004. High-intensity plasma ion nitriding of orthopedic materials: part I. Tribological study. Surf. Coat. Technol. 186: 305–313. https://doi.org/10.1016/j.surfcoat.2004.02.052.

[38] Oliver, W., R. Hutchings and J. Pethica. 1984. The wear behavior of nitrogen-implanted metals. Metall. Trans. A. 15: 2221–2229.

[39] Rautray, T. R., R. Narayanan and K. H. Kim. 2011. Ion implantation of titanium based biomaterials. Prog. Mater. Sci. 56: 1137–1177.

[40] Cengiz, S., Y. Azakli, M. Tarakci, L. Stanciu and Y. Gencer. 2017. Microarc oxidation discharge types and bio properties of the coating synthesized on zirconium. Mater. Sci. Eng. C. Mater. Biol. Appl. 77. 374–303.

[41] Suminov, I. V., P. N. Belkin., A. V. Apelfeld, V. B. Ludin., B. L. Krit and A. M. Borisov. 2011. Plasma electrolytic modification of metals and alloys. Moscow Technosfera. 464. ISBN 978-5-94836-266-3 (in Russian).

[42] Suminov, I. V., A. V. Apelfeld, V. B. Ludin, B. L. Krit and A. M. Borisov. 2005. Microarc oxidation theory, technology and equipment. Moscow Ecomet. 464–368. ISBN 5-89594-110-9 (in Russian).

[43] Yerokhin, A. L., X. Nie, A. Leyland, A. Matthews and S. J. Dowey. 1999. Plasma electrolysis for surface engineering. Surf. and Coat. Tech. 122: 73–93.

[44] Thompson, G. E., K. Shimizu and G. C. Wood. 1980. Observation of flaws in anodic films on aluminium. Nature 286: 471–472.

[45] O'Sullivan, J. P. and G. C. Wood. 1970. The morphology and mechanism of formation of porous anodic films on aluminium. Proc. Roy. Soc. (L) 317: 511–543.

[46] Neil, W. 1958. The preparation of cadmium niobate by an anodic spark reaction. J. Electrochem. Soc. 105: 544–547.

[47] Gruss, L. L. and W. Neil. 1963. Anodic spark reaction products in aluminate, tungstate and silicate solutions. Electrochem. Tech. 1: 283–287.

[48] Neil, W. and L. L. Gruss. 1963. Anodic film growth by anion deposition in aluminate, tungstate and phosphate solutions. J. Electrochem. Soc. 110: 853–855.

[49] Neil, W., L. L. Gruss and D. G. Husted. 1965. The anodic synthesis of CdS films. J. Electrochem. Soc. 112: 713–715.

[50] Mcneill, W. 1972. Electrolytic protective coating for magnesium. U.S. Patent #3791942.

[51] Mcneill, W. and L. L. Gruss. 1963. Anodic spark reaction process and articles. U.S. Patent #309580.

[52] Rakoch, A. G., A. V. Dub and A. A. Gladkova. 2012. Anodizing of light alloys under various electrical modes. Plasma electrolytic nanotechnology. M.: Staraya Basmannaya. 496. ISBN 978-5-904043-82-7 (in Russian).

[53] Rakoch, A. G., I. V. Bardin, V. L. Kovalev and A. G. Seferyan. 2013. Microarc oxidation of light constructional alloys: Part 2. Influence of the current waveform on the growth kinetics of microarc coatings on the surface of light construction alloys in alkali (pH ≤ 12.5) electrolytes. Rus. J. Non-Ferr. Met. 54: 345–348.

[54] Janaina, S. S., G. L. Sherlan, N. G Wesley, M. B. Odemir and C. P. Ernesto. 2014. Characterization of electrical discharges during spark anodization of zirconium in different electrolytes. Electrochim. Acta 130: 477–487.

[55] Hussein, R. O., X. Nie, D. O. Northwood, A. Yerokhin and A. Matthews. 2010. Spectroscopic study of electrolytic plasma and discharging behaviour during the plasma electrolytic oxidation (PEO) process. J. Phys. D: App. Phys. 43: 105–203.

[56] Rakoch, A. G., A. A. Gladkova, L. Zayar and D. M. Strekalina. 2015. The evidence of cathodic microdischarges during plasma electrolytic oxidation of light metallic alloys and micro-discharge intensity depending on pH of the electrolyte. Surf. Coat. Technol. 269: 138–144.

[57] Hartjen, P., A. Hoffmann, A. Henningsen, M. Barbeck, A. Kopp, L. Kluwe et al. 2018. Plasma electrolytic oxidation of titanium implant surfaces: microgroove-structures improve cellular adhesion and viability. *In Vivo* 32: 241–247.

[58] Yerokhin, A. L., A. Shatrov, V. Samsonov, P. Shashkov, A. Pilkington, A. Leyland et al. 2005. Oxide ceramic coatings on aluminium alloys produced by a pulsed bipolar plasma electrolytic oxidation process. Surf. Coat. Technol. 199: 150–157.

[59] Dunleavy, C. S., I. O. Golosnoy, J. A. Curran and T. W. Clyne. 2009. Characterisation of discharge events during plasma electrolytic oxidation. Surf. Coat. Technol. 203: 3410–3419.

[60] Yerokhin, A. L., A. L. Snisko, N. L. Gurevina, A. Leyland, A. Pilkington and A. Matthews. 2003. Discharge characterization in plasma electrolytic oxidation of aluminium. J. Phys. D: App. Phys. 36: 2110–2120.

[61] Rakoch, A. G., A. A. Gladkova and A. V. Dub. 2017. Plasma electrolytic oxidation of aluminum and titanium alloys. M.: MISiS. 170. ISBN 978-5-906846-518 (in Russian).

[62] Hussein, R. O., X. Nie and D. O. Northwood. 2013. Production of anti-corrosion coatings on light alloys (Al, Mg, Ti) by plasma-electrolytic oxidation (PEO). Developments in Corrosion Protection. Chapter 11.

[63] Hussein, R. O., X. Nie and D. O. Northwood. 2013. An investigation of ceramic coating growth mechanisms in plasma electrolytic oxidation (PEO) processing. Electrochim. Acta 112: 111–119.

[64] Gencer, Y. and A. E. Gulec. 2012. The effect of Zn on the microarc oxidation coating behavior of synthetic Al–Zn binary alloys. J. Alloys Compd. 525: 159–165.

[65] Gladkova, A. A., V. V. Khovaylo, A. G. Rakoch, N. A. Predein, P. V. Truong, H. Kosukegawa et al. 2016. Proceedings of XVI International Symposium on Advanced Fluid Information. 140–142.

[66] Oh, Y. -J., J. -I. Mun and J. -H. Kim. 2009. Effects of alloying elements on microstructure and protective properties of Al_2O_3 coatings formed on aluminum alloy substrates by plasma electrolysis. Surf. Coat. Technol. 204: 141–148.

[67] Sah, S. P., E. Tsuji, Y. Aoki and H. Habazaki. 2012. Cathodic pulse breakdown of anodic films on aluminium in alkaline silicate electrolyte: Understanding the role of cathodic half-cycle in AC plasma electrolytic oxidation. Corros. Sci. 55: 90–96.

[68] Arrabal, R., E. Matykina, T. Hashimoto, P. Skeldon and G. E. Thompson. 2009. Characterization of AC PEO coatings on magnesium alloys. Surf. Coat. Technol. 203: 2207–2220.

[69] Rokosz, K., T. Hryniewicz and S. Gaiaschi. 2018. Novel porous phosphorus-calcium-magnesium coatings on titanium with copper or zinc obtained by DC plasma electrolytic oxidation: fabrication and characterization. Materials (Basel) 11: 1680.

[70] Toorani, M., M. Aliofkhazraei and R. Naderi. 2019. Ceria-embedded MAO process as pretreatment for corrosion protection of epoxy films applied on AZ31-magnesium alloy. J. Alloys & Comp. 785: 669–683.

[71] Lu, X., C. Blawert, K. U. Kainer, T. Zhang and L. Z. Mikhail. 2018. Influence of particle additions on corrosion and wear resistance of plasma electrolytic oxidation coatings on Mg alloy. Surf. Coat. Technol. 352: 1–14.

[72] Krishtal, M. M., P. V. Ivashin, A. V. Polunin and E. D. Borgardt. 2019. The effect of dispersity of silicon dioxide nanoparticles added to electrolyte on the composition and properties of oxide layers formed by plasma electrolytic oxidation on magnesium 9995A. Mat. Letters. 241: 119–122.

[73] Tran, Q. -P., T. -S. Chin, Y. -C. Kuo, C. -X. Jin and D. Q. Dang. 2018. Diamond powder incorporated oxide layers formed on 6061 Al alloy by plasma electrolytic oxidation. J. Alloys & Comp. 751: 289–298.

[74] An, L., Y. Ma, Y. Liu, L. Sun and Z. Wang. 2018. Effects of additives, voltage and their interactions on PEO coatings formed on magnesium alloys. Surf. Coat. Technol. 354: 226–235.

[75] Lou, B. -S., J. -W. Lee, C. -M. Tseng, Y. -Y. Lin and C. -A. Yen. 2018. Mechanical property and corrosion resistance evaluation of AZ31 magnesium alloys by plasma electrolytic oxidation treatment: Effect of MoS_2 particle addition. Surf. Coat. Technol. 350: 813–822.

[76] Peitao, G., T. Mingyang and Z. Chaoyang. 2019. Tribological and corrosion resistance properties of graphite composite coating on AZ31 Mg alloy surface produced by plasma electrolytic oxidation. Surf. Coat. Technol. 359: 197–205.

[77] Liu, W., Y. Liu, Y. Lin, Z. Zhang, S. Feng, T. Shi et al. 2019. Effects of graphene on structure and corrosion resistance of plasma electrolytic oxidation coatings formed on D16T Al alloy. App. Surf. Sci. 475: 645–659.

[78] Stojadinović, S., N. Tadić, N. Radić, B. Grbić and R. Vasilić. 2018. CdS particles modified TiO_2 coatings formed by plasma electrolytic oxidation with enhanced photocatalytic activity. Surf. Coat. Technol. 344: 528–533.

[79] Chen, L., Y. Qu, X. Yang, B. Liao and W. Cheng. 2017. Characterization and first-principles calculations of WO_3/TiO_2 composite films on titanium prepared by microarc oxidation. Materials Chemistry and Physics 201: 311–322.

[80] Ebrahimi, S., A. B. Khiabani, B. Yarmand and M. A. Asghari. 2019. Improving optoelectrical properties of photoactive anatase TiO_2 coating using rGO incorporation during plasma electrolytic oxidation. Ceramics Intl. 45: 1746–1754.

[81] Friedemann, E. R., K. Thiel, T. M. Gesing and P. Plagemann. 2018. Photocatalytic activity of TiO_2 layers produced with plasma electrolytic oxidation. Surf. Coat. Technol. 344: 710–721.

[82] Stojadinović, S. and R. Vasilić. 2019. Photoluminescence properties of Er^{3+}/Yb^{3+} doped ZrO_2 coatings formed by plasma electrolytic oxidation. J. Luminesc. 208: 296–301.

[83] Stojadinović, S., N. Tadić and R. Vasilić. 2018. Eu^{2+} photoluminescence in Al_2O_3 coatings obtained by plasma electrolytic oxidation. J. Luminesc. 199: 240–244.

[84] Ma, C., D. Cheng, Y. Zhu, Z. Yan and S. Zheng. 2018. Investigation of a self-lubricating coating for diesel engine pistons, as produced by combined microarc oxidation and electrophoresis. Wear. 394–395: 109–112.

[85] Matykina, E., R. Arrabal, M. Mohedano, B. Mingo and M. C. Merino. 2017. Recent advances in energy efficient PEO processing of aluminium alloys. Transactions of Nonferrous Metals Society of China 27: 1439–1454.

[86] Lim, Y. W., S. Y. Kwon, D. H. Sun, H. E. Kim and Y. S. Kim. 2009. Enhanced cell integration to titanium alloy by surface treatment with microarc oxidation: a pilot study. Clin. Orthop. Relat. Res. 467: 2251–2258.

[87] Marques, I. D., V. A. Barão, N. C. da Cruz, J. C. -C. Yuan, M. F. Mesquita, A. P. R. Filho et al. 2015. Electrochemical behavior of bioactive coatings on cp-Ti surface for dental application. Corr. Sci. 100: 133–146.

[88] Bai, Y., R. Zhou, J. Cao, D. Wei, Q. Du, B. Li et al. 2017. Microarc oxidation coating covered Ti implants with micro-scale gouges formed by a multi-step treatment for improving osseointegration. Mater. Sci. Eng. C. Mater. Biol. Appl. 76: 908–917.

[89] Qiao, L., Z. Ding, L. Zhang and T. Niu. 2013. Microarc oxidation of titanium surfaces on osteoblast morphology and cytoskeleton. Hua Xi Kou Qiang Yi Xue Za Zhi. 31: 468–71.

[90] Shi, L., L. Wang, Y. Zhang, Z. Guo, Z. X. Wu, D. Liu et al. 2012. Improving fixation strength of pedicle screw by microarc oxidation treatment: an experimental study of osteoporotic spine in sheep. J. Orthop. Res. 30: 1296–303.

[91] Li, L. H., H. W. Kim, S. H. Lee, Y. M. Kong and H. E. Kim. 2005. Biocompatibility of titanium implants modified by microarc oxidation and hydroxyapatite coating. J. Biomed. Mater. Res. A. 73: 48–54.

[92] Li, G., H. Cao, W. Zhang, X. Ding, G. Yang, Y. Qiao et al. 2016. Enhanced osseointegration of hierarchical micro/nanotopographic titanium fabricated by microarc oxidation and electrochemical treatment. ACS Appl. Mater. Interfaces 8: 3840–3852.

[93] Deng, Z., L. Wang, D. Zhang, J. Liu, C. Liu and J. Ma. 2014. Lanthanum-containing hydroxyapatite coating on ultrafine-grained titanium by micro-arc oxidation: a promising strategy to enhance overall performance of titanium. Med. Sci. Monit. 20: 163–6.

[94] Rendon, M. E., V. Duque, D. Quintero, S. M. Robledo, M. C. Harmsen and F. Echeverria. 2018. Improved corrosion resistance of commercially pure magnesium after its modification by plasma electrolytic oxidation with organic additives. J. Biomater. Appl. 33: 725–740.

[95] Xu, L., K. Zhang, C. Wu, X. Lei, J. Ding, X. Shi et al. 2017. Micro-arc oxidation enhances the blood compatibility of ultrafine-grained pure titanium. Materials (Basel) 10: 1446.

[96] Lim, Y. W., S. Y. Kwon, D. H. Sun and Y. S. Kim. 2011. The Otto Aufranc Award: enhanced biocompatibility of stainless steel implants by titanium coating and microarc oxidation. Clin. Orthop. Relat. Res. 469: 330–338.

[97] Sowa, M. and W. Simka. 2018. Effect of DC plasma electrolytic oxidation on surface characteristics and corrosion resistance of zirconium. Materials (Basel) 11: 723. doi:10.3390/ma11050723.

[98] Goularte, M. A., G. F. Barbosa, N. C. da Cruz and L. M. Hirakata. 2016. Achieving surface chemical and morphologic alterations on tantalum by plasma electrolytic oxidation. Int. J. Implant. Dent. 2: 12.

[99] White, L., Y. Koo, S. Neralla, J. Sankar and Y. Yun. 2016. Enhanced mechanical properties and increased corrosion resistance of a biodegradable magnesium alloy by plasma electrolytic oxidation (PEO). Mater. Sci. Eng. B. Solid State Mater. Adv. Technol. 208: 39–46.

[100] Wu, Y. F., Y. M. Wang, Y. B. Jing, J. P. Zhuang, J. L. Yan, Z. K. Shao et al. 2017. *In vivo* study of microarc oxidation coated biodegradable magnesium plate to heal bone fracture defect of 3 mm width. Colloids Surf. B. Biointerfaces 158: 147–156.

[101] Rendon, M. E., V. Duque, D. Quintero, S. M. Robledo, M. C. Harmsen and F. Echeverria. 2018. Improved corrosion resistance of commercially pure magnesium after its modification by plasma electrolytic oxidation with organic additives. J. Biomater. Appl. 33: 725–740.

[102] Anawati, A., H. Asoh and S. Ono. 2016. Effects of alloying element Ca on the corrosion behavior and bioactivity of anodic films formed on AM60 Mg alloys. Materials (Basel) 10: 11.

[103] Tian, P., F. Peng, D. Wang and X. Liu. 2016. Corrosion behavior and cytocompatibility of fluoride-incorporated plasma electrolytic oxidation coating on biodegradable AZ31 alloy. Regen. Biomater. 4: 1–10.

[104] Rokosz, K., T. Hryniewicz, S. Gaiaschi, P. Chapon, S. Raaen, W. Malorny et al. 2018. Development of porous coatings enriched with magnesium and zinc obtained by DC plasma electrolytic oxidation. Micromachines (Basel) 9: 332.

[105] Zeng, R. C., L. Y. Cui, K. Jiang, R. Liu, B. D. Zhao and Y. F. Zheng. 2016. *In vitro* corrosion and cytocompatibility of a microarc oxidation coating and poly(L-lactic acid) composite coating on Mg-1Li-1Ca alloy for orthopedic implants. ACS Appl. Mater. Interfaces 8: 10014–10028.

[106] Peng, F., D. Wang, Y. Tian, H. Cao, Y. Qiao and X. Liu. 2017. Sealing the pores of PEO coating with Mg-Al layered double hydroxide: enhanced corrosion resistance, cytocompatibility and drug delivery ability. Sci Rep. 7: 8167.

[107] Zoubi, W. A., J. H. Min and Y. G. Ko. 2017. Hybrid organic-inorganic coatings via electron transfer behaviour. Sci. Rep. 7: 7063.

[108] Bottagisio, M., A. B. Lovati, F. Galbusera, L. Drago and G. Banfi. 2019. A precautionary approach to guide the use of transition metal-based nanotechnology to prevent orthopedic infections. Materials (Basel) 12: 314.

[109] Beyth, N., Y. H. Haddad, A. Domb, W. Khan and R. Hazan. 2015. Alternative antimicrobial approach: Nano-antimicrobial materials. Evid. Based Complement. Alternat. Med. 2015: 246012.

[110] Gottenbos, B., D. W. Grijpma, H. C. van der Mei, J. Feijen and H. J. Busscher. 2001. Antimicrobial effects of positively charged surfaces on adhering Gram-positive and Gram-negative bacteria. J. Antimicrob. Chemother. 48: 7–13.

[111] Markowska, K., A. M. Grudniak and K. I. Wolska. 2013. Silver nanoparticles as an alternative strategy against bacterial biofilms. Acta Biochim. Pol. 60: 523–530.

[112] Yu, Q., J. Li, Y. Zhang, Y. Wang, L. Liu and M. Li. 2016. Inhibition of gold nanoparticles (Au NPs) on pathogenic biofilm formation and invasion to host cells. Sci. Rep. 6: 26667.

[113] Miao, L., C. Wang, J. Hou, P. Wang, Y. Ao, Y. Li et al. 2016. Aggregation and removal of copper oxide (CuO) nanoparticles in wastewater environment and their effects on the microbial activities of wastewater biofilms. Bioresour. Technol. 216: 537–544.

[114] Clarke, I. C., D. D. Green, G. Pezzoti and D. Donaldson. 2005. 20 year experience of zirconia total hip replacements. pp. 67–78. *In:* D'Antonio, J. A. and M. Dietrich (eds.). [1st]. Bioceramics and Alternative Bearings in Joint Arthroplasty. Steinkopff, Dresden, Germany. ISBN 978-3-7985-1518-5.

[115] Margabandhu, M., S. Sendhilnathan, S. Maragathavalli, V. Karthikeyan and B. Annadurai. 2015. Synthesis characterization and antibacterial activity of iron oxide nanoparticles. Glob. J. Bio Sci. Biotechnol. 4: 335.

[116] Itabashi, T., K. Narita, A. Ono, K. Wada, T. Tanaka, G. Kumagai et al. 2017. Bactericidal and antimicrobial effects of pure titanium and titanium alloy treated with short-term, low-energy UV irradiation. Bone Jt. Res. 6: 108–112.

[117] Sirelkhatim, A., S. Mahmud, A. Seeni, N. H. M. Kaus, L. C. Ann, S. K. M. Bakhori et al. 2015. Review on zinc oxide nanoparticles: antibacterial activity and toxicity mechanism. Nanomicro Lett. 7: 219–242.

Index

Milton Keynes UK
Ingram Content Group UK Ltd.
UKHW051015071024
449327UK00012B/261